T0140537

Studies in Computational Intelligence

Volume 529

Series Editor

Janusz Kacprzyk, Polish Academy of Sciences, Warsaw, Poland
e-mail: kacprzyk@ibspan.waw.pl

For further volumes:
http://www.springer.com/series/7092

About this Series

The series "Studies in Computational Intelligence" (SCI) publishes new developments and advances in the various areas of computational intelligence—quickly and with a high quality. The intent is to cover the theory, applications, and design methods of computational intelligence, as embedded in the fields of engineering, computer science, physics and life sciences, as well as the methodologies behind them. The series contains monographs, lecture notes and edited volumes in computational intelligence spanning the areas of neural networks, connectionist systems, genetic algorithms, evolutionary computation, artificial intelligence, cellular automata, self-organizing systems, soft computing, fuzzy systems, and hybrid intelligent systems. Of particular value to both the contributors and the readership are the short publication timeframe and the world-wide distribution, which enable both wide and rapid dissemination of research output.

Nadia Nedjah · Luiza de Macedo Mourelle

Hardware for Soft Computing and Soft Computing for Hardware

 Springer

Nadia Nedjah
Department of Electronics Engineering
  and Telecommunications
Faculty of Engineering
State University of Rio de Janeiro
Rio de Janeiro
Brazil

Luiza de Macedo Mourelle
Department of Systems Engineering
  and Computation
Faculty of Engineering
State University of Rio de Janeiro
Rio de Janeiro
Brazil

ISSN 1860-949X
ISSN 1860-9503   (electronic)
ISBN 978-3-319-37874-9
ISBN 978-3-319-03110-1   (eBook)
DOI 10.1007/978-3-319-03110-1
Springer Cham Heidelberg New York Dordrecht London

Printed on acid-free paper

Springer is part of Springer Science+Business Media (www.springer.com)

Preface

Evolutionary and Genetic Algorithms, Artificial Neural Networks, Fuzzy Systems, Particle Swarm and Ant colony Optimization are becoming omnipresent in almost every intelligent system design. Just to name few, engineering, control, economics and forecasting are some of the scientific fields that enjoy the use of these techniques to solve real-world engineering problems. Unfortunately, the majority of the applications is complex and so requires a large computational effort to yield useful and practical results. Therefore, dedicated hardware for evolutionary, neural and fuzzy computation is becoming a key issue for designers. With the spread of reconfigurable hardware such as FPGAs, digital as well as analog hardware implementations of such computation become cost-effective.

Nowadays, embedded systems are omnipresent in homes, cars, cell phones, medical instruments etc. Embedded systems specifications usually impose very tight constraints with respect to cost, response time and power consumption, among other characteristics. They often include software, hardware and interfacing subsystems. Furthermore, the design of these kind of systems often requires concurrent optimization of several design objectives, which are conflicting in most of the cases. Also, these project call for the resolution of NP-hard problems. Their diversity and complexity require new design methodologies.

The idea behind this book is to offer a variety of hardware designs for soft computing techniques that can be embedded in any final product. Also, to introduce the successful application of soft computing technique to solve many hard problem encountered during the design of embedded hardware designs.

Part I: Hardware for Soft Computing

In Chapter 1, which is entitled *A Reconfigurable Hardware for Genetic Algorithms*, we propose a massively parallel architecture of a hardware implementation of genetic algorithms. This design is quite innovative as it provides a viable solution to the fitness computation problem, which depends heavily on the problem-specific knowledge. The proposed architecture is completely independent of such specifics. It implements the fitness computation using a neural network.

Chapter 2, which is entitled *Genetic Algorithms on Network-on-Chip*, presents a parallel implementation on a network-on-chip platform of the genetic algorithm. The implementation is based on the island model, in which serial isolated subpopulations evolve in parallel and each one is controlled by a single processor. Rubem Euzébio Ferreira, M.Sc. collaborated in the development of this chapter.

Chapter 3, which is entitled *A Reconfigurable Hardware for Particle Swarm Optimization*, reports on a novel massively parallel co-processor for PSO implemented using reconfigurable hardware. The implementation results show that the proposed architecture is very promising as it achieved superior performance in terms of execution time when compared to the direct software execution of the algorithm. Rogério de Moraes Calazan, M.Sc. collaborated in the development of this chapter.

Chapter 4, which is entitled *Particle Swarm Optimization on Crossbar based MP-SoC*, investigates the performance characteristics of a parallel application running on this platform we based the interconnection network in the crossbar topology. In this kind of interconnection, processors have full access to their own memory module simultaneously. This chapter also details the specification and modeling of an interconnection network based on crossbar topology. Fábio Gonçalves Pessanha, M.Sc. collaborated in the development of this chapter.

Chapter 5, which is entitled *A Reconfigurable Hardware for Artificial Neural Networks*, devises a hardware architecture for ANNs that takes advantage of the dedicated adder blocks, commonly called MACs to compute both the weighted sum and the activation function. The proposed architecture requires a reduced silicon area considering the fact that the MACs come for free as these are FPGA's built-in cores. The hardware is as fast as existing ones as it is massively parallel. Besides, the proposed hardware can adjust itself on-the-fly to the user-defined topology of the neural network, with no extra configuration, which is a very nice characteristic in robot-like systems considering the possibility of the same hardware may be exploited in different tasks. Rodrigo Matins da Silva, M.Sc. collaborated in the development of this chapter.

Chapter 6, which is entitled *A Reconfigurable Hardware for Fuzzy Controllers*, elaborates on the development of a reconfigurable efficient architecture for fuzzy controllers, suitable for embedding. The architecture is parameterizable so it allows the setup and configuration of the controller so it can be used for various problem applications. An application of fuzzy controllers was implemented and its cost and performance are presented. Paulo Renato S. S. Sandres, M.Sc. collaborated in the development of this chapter.

Chapter 7, which is entitled *A Reconfigurable Hardware for Subtractive Clustering*, presents the development of a reconfigurable hardware for classification system of radioactive elements with a fast and efficient response. To achieve this goal, the hardware implementation of subtractive clustering algorithm is proposed. The hardware is generic, so it can be used in many problems of data classification, omnipresent in identification systems. Marcos Sanatana Farias, M.Sc. collaborated in the development of this chapter.

Chapter 8, which is entitled *A Reconfigurable Hardware for DNA Matching*, proposes a novel parallel hardware architecture for DNA matching based on the steps of the BLAST algorithm. The design is scalable so that its structure can be adjusted depending on size of the subject and query DNA sequences. Moreover, the number of units used to perform in parallel can also be scaled depending some characteristics of the algorithm. The design was synthesized and programmed into FPGA. The trade-off between cost and performance were analyzed to evaluate different design configuration. Edgar José Garcia Neto Segundo, M.Sc. collaborated in the development of this chapter.

Part II: Soft Computing for Hardware

Chapter 9, which is entitled *Synchronous Finite State Machines Design with Quantum-inspired Evolutionary Computation*, explores an evolutionary methodology based on the principles of quantum computing to synthesize finite state machines. First, we optimally solve the state assignment NP-complete problem, which is inherent to designing any synchronous finite state machines. This is motivated by the fact that with an optimal state assignment, one can physically implement the state machine in question using a minimal hardware area and response time. Second, with the optimal state assignment provided, we propose to use the same evolutionary methodology to yield an optimal evolutionary hardware that implements the state machine control component. The evolved hardware requires a minimal hardware area and imposes a minimal propagation delay on the machine output signals. Marcos Paulo Araujo Melo, M.Sc. collaborated in the development of this chapter.

Chapter 10, which is entitled *Application Mapping in Network-on-Chip using Evolutionary Multi-objective Optimization*, uses multi-objective evolutionary optimization to address the problem of mapping topologically pre-selected sets IPs, which constitute the set of optimal solutions that were found for the IP assignment problem, on the tiles of a mesh-based NoC. The IP mapping optimization is driven by the area occupied, execution time and power consumption. Marcus Vinícius Carvalho da Silva, M.Sc. collaborated in the development of this chapter.

Chapter 11, which is entitled *Application Routing in Network-on-Chip using Ant Colony Optimization*, takes advantage of ant colony algorithms to find and optimize routes in a mesh-based NoC, where several randomly generated applications have been mapped. The routing optimization is driven by the minimization of total latency in packets transmission between tasks. The simulation results show the effectiveness of the ant colony inspired routing by comparing it with general purpose algorithms for deadlock free routing. Luneque Del Rio de Souza e Silva Júnior, M.Sc. collaborated in the development of this chapter.

Acknowledgments

First o all, we would like to acknowledge the help of all the mentioned collaborators, as mentioned before, in the elaboration of the chapters of this book. We are very much grateful to the editors would also like to thank Prof. Janusz Kacprzyk, the editor-in-chief of the Studies in Computational Intelligence Book Series and Dr. Thomas Ditzinger from Springer-Verlag, Germany for their editorial assistance and excellent collaboration to produce this scientific work. We hope that the reader will share our excitement on this volume and will find it useful.

We are grateful to FAPERJ (*Fundação de Amparo à Pesquisa do Estado do Rio de Janeiro*, www.faperj.br), CNPq (*Conselho Nacional de Desenvolvimento Científico e Tecnológico*, www.cnpq.br) and CAPES (*Coordenação de Aperfeiçoamento de Pessoal de Ensino Superior*, www.capes.gov.br) for their continuous financial support.

Rio de Janeiro, August 2013

Nadia Nedjah
Department of Electronics Engineering and Telecommunications
Faculty of Engineering
State University of Rio de Janeiro, Brazil

Luiza de Macedo Mourelle
Department of Systems Engineering and Computation
Faculty of Engineering
State University of Rio de Janeiro, Brazil

Contents

Part I
Hardware for Soft Computing

Chapter 1
A Reconfigurable Hardware for Genetic Algorithms

Abstract. In this chapter, we propose a massively parallel architecture of a hardware implementation of genetic algorithms. This design is quite innovative as it provides a viable solution to the fitness computation problem, which depends heavily on the problem-specific knowledge. The proposed architecture is completely independent of such specifics. It implements the fitness computation using a neural network. The hardware implementation of the used neural network is stochastic and thus minimises the required hardware area without much increase in response time. Last but not least, we demonstrate the characteristics of the proposed hardware and compare it to existing ones.

1.1 Introduction

Generally speaking, a *genetic algorithm* is a process that evolves a set of *individuals*, also called *chromosomes*, which constitutes the *generational population*, producing a new population. The individuals represent a solution to the problem in consideration. The freshly produced population is yield using some genetic operators such as *selection*, *crossover* and *mutation* that attempt to simulate the natural breeding process in the hope of generating new solutions that are *fitter*, i.e. adhere more the problem constraints.

Previous work on hardware genetic algorithms can be found in [5, 10, 12]. Mainly, Earlier designs are hardware/software codesigns and they can be divided into three distinct categories: *(i)* those that implement the fitness computation in hardware and all the remaining steps including the genetic operators in software, claiming that the bulk computation within genetic evolution is the fitness computation. The hardware is problem-dependent; *(ii)* and those that implement the fitness computation in software and the rest in hardware, claiming that the ideal candidate are the genetic operators as these exhibit regularity and generality [2, 7]. *(iii)* those that implement the whole genetic algorithm in hardware [10]. We believe that both approaches are worthwhile but a hardware-only implementation of both the fitness calculation and genetic operators is also valuable. Furthermore, a hardware implementation that is problem-independent is yet more useful.

N. Nedjah and L. de Macedo Mourelle, *Hardware for Soft Computing and Soft Computing for Hardware*, Studies in Computational Intelligence 529,
DOI: 10.1007/978-3-319-03110-1_1, © Springer International Publishing Switzerland 2014

The remainder of this chapter is divided into five sections. In Section 1.2, we describe the principles of genetic algorithms. Subsequently, in Section 1.3, we propose and describe the overall hardware architecture of the problem-independent genetic algorithm. Thereafter, in Section 1.4, we detail the architecture of each of the component included in the hardware genetic algorithm proposed. Then, in Section 1.5, assess the performance of the proposed architecture. Finally, we draw some conclusions 9.9.

1.2   Principles of Genetic Algorithms

Genetic algorithms maintain a *population* of *individuals* that evolve according to *selection rules* and other *genetic operators*, such as *mutation* and *crossover*. Each individual receives a measure of *fitness*. Selection focuses on high fitness individuals. Mutation and crossover provide general heuristics that simulate the reproduction process. Those operators attempt to perturb the characteristics of the parent individuals as to generate distinct offspring individuals.

Genetic algorithms are implemented through the procedure described by Algorithm 1.1, wherein parameters *ps*, *ef* and *gn* are the population size, the expected fitness of the returned solution and the maximum number of generation allowed respectively.

Algorithm 1.1. GA – Genetic algorithms basic cycle

Require: population size *ps*, expected fitness *ef*, generation number *gn*
Ensure: the problem solution
  generation := 0
  population := initialPopulation()
  fitness := evaluate(population)
  repeat
    parents := select(population)
    population := mutate(crossover(parents))
    fitness := evaluate(population)
    generation := generation + 1
  until (fitness[i] = *ef*, $1 \leq i \leq ps$) OR (generation $\geq gn$)

In Algorithm 1.1, function *intialPopulation* returns a valid random set of individuals that compose the population of first generation while function *evaluate* returns the fitness of a given population storing the result into fitness. Function *select* chooses according to some random criterion that privilege fitter individuals, the individuals that should be used to generate the population of the next generation and function *crossover* and *mutate* implement the crossover and mutation process respectively to actually yield the new population.

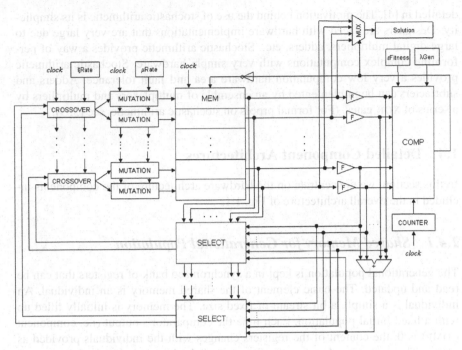

Fig. 1.1 Overall architecture of the hardware genetic algorithm proposed

1.3 Overall Architecture for the Hardware Genetic Algorithm

Clearly, for hardware genetic algorithms, individuals are always represented using their binary representation. Almost all aspects of genetic algorithms are very attractive for hardware implementation. The selection, crossover and mutation processes are generic and so are problem-independent. The main issue in the hardware implementation of genetic algorithms is the computation of individual's fitness values. This computation depends on problem-specific knowledge. The novel contribution of the work consists of using neural network hardware to compute the fitness of individuals. The software version of the neural network is trained with a variety of individual examples. Using a hardware neural network to compute individual fitness yields a hardware genetic algorithm that is fully problem-independent.

The overall architecture of the proposed hardware is given Fig. 1.1. It is massively parallel. The selection process is performed in one clock cycle while the crossover and mutation processes are completed within two clock cycles.

The fitness of individual in the generational population is evaluated using hardware neural networks, which take advantage of stochastic representation of signals to reduce the hardware area required [9]. Stochastic computing principles are well detailed in [4]. The motivation behind the use of stochastic arithmetic is its simplicity. Designers are faced with hardware implementations that are very large due to large digital multipliers, adders, etc.. Stochastic arithmetic provides a way of performing complex computations with very simple hardware. Stochastic arithmetic provides a very low computation hardware area and fault tolerance. Adders and subtracters can be implemented by an ensemble of multiplexers and multipliers by a series of XOR gates. (For formal proofs on stochastic arithmetic, see [3, 9].)

1.4 Detailed Component Architectures

In this section, we concentrate on the hardware architecture of the components included in the overall architecture of Fig. 1.1.

1.4.1 Shared Memory for Generational Population

The generational population is kept in a synchronised bank of registers that can be read and updated. The basic element of the shared memory is an individual. An individual is a simply a bit stream of fixed size. The memory is initially filled up with a fixed initial population. Each time the comparator's output (i.e. component COMP) is 0, the content of the registers changes with the individuals provided as inputs. This happens whenever the fitness f of the best individual of the current generation is not as expected (i.e., $f > \varepsilon Fitness$) and the current generation g is not the last allowed one (i.e., $g \neq \lambda Gen$). Note that $\varepsilon Fitness$ and λGen are two registers that store the expected fitness value and the maximum number of generation allowed respectively.

1.4.2 Random Number Generator

A central component to the proposed hardware architecture of genetic algorithms is a source of pseudorandom noise. A source of pseudorandom digital noise consists of a *linear feedback shift register* or LFSR, described by first in [1] and by many others, for instance [3], LFSRs are very practical as they can easily be constructed using standard digital components.

Linear feedback shift registers can be implemented in two ways. The Fibonacci implementation consists of a simple shift register in which a binary-weighted modulo-2 sum of the taps is fed back to the input. Recall that modulo-2 sum of two one-bit binary numbers yields 0 if the two numbers are identical and 1 if not. The Galois implementation consists of a shift register, the content of which is modified at every step by a binary-weighted value of the output stage. The architecture of the LFSR using these methods are shown in Fig. 1.2.

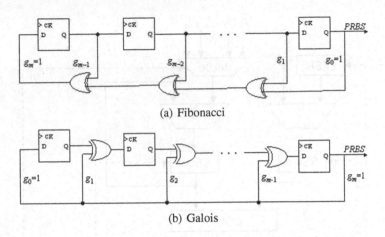

(a) Fibonacci

(b) Galois

Fig. 1.2 Pseudorandom bitstream generators - Fibonacci vs. Galois implementation

Left feedback shift registers such as those of Fig. 1.2 can be used to generate multiple pseudorandom bit sequences. However, the taps from which these sequences are yield as well as the length of the LFSR must be carefully chosen. (See [3, 4] for possible length/tap position choices).

1.4.3 Selection Component

The selection component implements a variation of the roulette wheel selection. The interface of this component consists of all the individuals, say $i_1, i_2, \ldots, i_{n-1}, i_n$ of the generational population of size n together with the respective fitness, say $f_1, f_2, \ldots, f_{n-1}, f_n$ and the overall sum of all these fitness values, say sum. The component proceeds as described in the following steps:

1. A random number, say ρ is generated;
2. The sum of the individual's fitness values is scaled down using ρ, i.e $ssum := sum - \rho$;
3. Choose an individual, say i_j from the selection pool and cumulate the corresponding fitness f_j, i.e. $csum := csum + f_j$;
4. Compare the scaled sum and the so far cumulated sum and select individual i_j if $csum > ssum$, otherwise go back to step 1;
5. When the first individual is selected, go back to step 1 and apply the same process to select the second individual.

The architecture of the selection component is shown in Fig. 1.3. The above iterative process is implemented using a state machine (CONTROLLER in Fig. 1.3). The state machine has 6 states and the associated state transition function is described in Fig. 1.4. The actions performed in each state of the controller machine are described

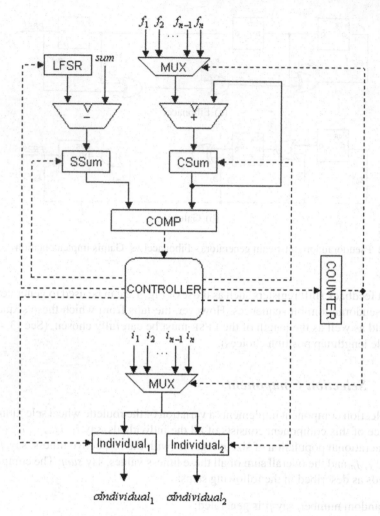

Fig. 1.3 The architecture of the selection component

below. Signal *compare* is set when the either an individual having the expected fitness is found or the last generation has passed.

S_0:   initialise counter;
        load register CSUM with 0;
S_1:   stop the random number generator;
S_2:   load register SSUM;
S_3:   load register CSUM;
S_4:   if *compare* = 1 then
            if *step* = 0 then
                load register INDIVIDUAL$_1$;

start the random number generator;
else load register INDIVIDUAL$_2$;
increment the counter;
S_5:   if *step* = 0 then set *step*;

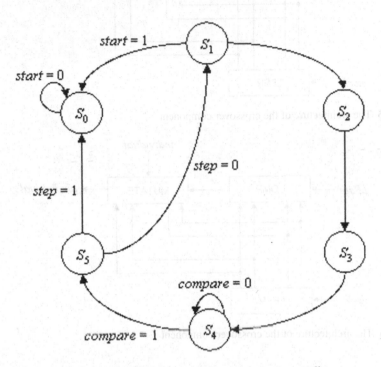

Fig. 1.4 The state transition function of the selection component controller

1.4.4   Genetic Operator's Components

The genetic operators are the crossover followed by the mutation. The crossover component implements the double-point crossover. It uses a linear feedback shift register which provides the random number that allows the component to decide whether to actually perform the crossover or not. This depends on whether the randomised number surpasses the informed crossover rate $\xi Rate$. In the case it does, the bits of the less significant half of the randomised number is used as the first crossover point and the most significant part as the second one.

The mutation component also uses a random number generator. The generated number must be bigger that the given mutation rate $\mu Rate$ for the mutation to occur. The bits of the randomised number are also used as way to choose the mutation

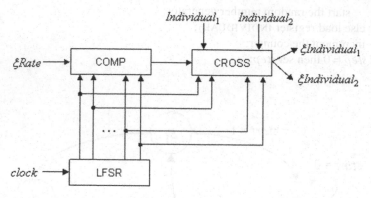

Fig. 1.5 The architecture of the crossover component

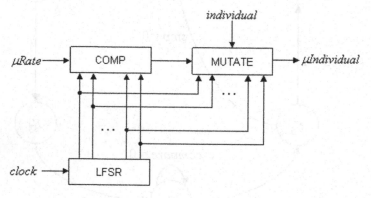

Fig. 1.6 The architecture of the crossover component

degree of the individual. Starting from the less significant bit of both the random number and the individual, if the bit in the former is 1 then the corresponding bit in the later is complemented and otherwise it is kept unchanged. The hardware architecture of the mutation component is given in Fig. 1.6.

1.4.5 Fitness Evaluation Component

The individual fitness measure is estimated using neural networks. In previous work, the authors proposed and implemented a hardware for neural networks [9]. The implementation uses stochastic signals and therefore reduces very significantly the hardware area required for the network. The network topology used is the fully-connected feed-forward. The neuron architecture is given in Fig. 1.7. (More details can be found in [9].) For the hardware genetic implementation, the number of input neurons is the same as the size of the individual. The output neuron are augmented with a shift register to store the final result. The training phase is supposed to be performed before the first use within the hardware genetic algorithm.

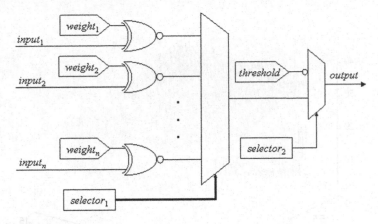

Fig. 1.7 Stochastic bipolar neuron architecture ([9])

1.5 Performance Results

The hardware genetic algorithm proposed was simulated then programmed into an Spartan3 Xilinx FPGA [14]. In order to assess the performance of the proposed hardware genetic algorithm, we maximise the function that was first used in [8]. It was also used by Scott, Seth and Samal to evaluate their hardware implementation for genetic algorithms [11]. The function is not easy to maximise, which is clear from the function plot of Fig. 1.8. The training phase of the neural network was done by software using toolbox offered in MatLab [6].

$$f(x,y) = 21.5 + x\sin(4\pi x) + y\sin(20\pi y),$$

$$-3.0 \le x_1 \le 12.1$$
$$4.1 \le x_2 \le 5.8$$

(1.1)

The characteristics of the software and hardware implementations proposed in [11] and those of the hardware genetic algorithm we proposed in this chapter are compared in Table 1.1. It is clear that the hardware implementations are both much faster than the software version. One can clearly note that our implementation (PHGA) requires more than twice that required by Scott, Seth and Samal's implementation (HEGA). Note, however, that the hardware area necessary to the computation of the fitness function is not included as it is not given in [11]. From another perspective, PHGA is more than five time faster as it is massively parallel. We also believe that the computation of the fitness function is much faster with the neural network. Observe that PHGA evolved a better solution.

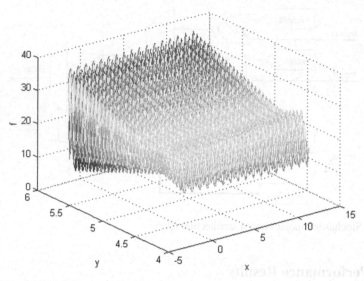

Fig. 1.8 Plotting Michalewics's function ([8])

Table 1.1 Comparison of the performance results: software genetic algorithms (SGA), hardware engine for genetic algorithms (HEGA) and proposed hardware genetic algorithms (PHGA). (The area is expressed in terms of CLBs and the time is in seconds.)

Implementation	time	area	solution	x	y	area× time
SGA	40600	0	38.5764	–	–	–
HEGA	972	870	38.8419	–	–	845640
PHGA	189	1884	38.8483	11.6241	5.7252	356076

1.6 Summary

In this chapter, we proposed a novel hardware architecture for genetic algorithms. It is novel in the sense that is massively parallel and problem-independent. It uses neural networks to compute the fitness measure. Of course, for each type of problem, the neuron weights need to be updated with those obtained in the training phase. Without any doubts, the proposed hardware is extremely faster than the software implementation. Furthermore, it is much faster than the hardware engine proposed by Scott, Seth and Samal in [11]. However, it seems that our implementation requires almost twice the hardware are used to implement their architecture. Nevertheless, we do not have an exact record of the hardware are consumed in [11] as the authors did not provide nor include the hardware required to implement the fitness module for Michalewics's function [8].

References

1. Bade, S.L., Hutchings, B.L.: FPGA-Based Stochastic Neural Networks - Implementation. In: IEEE Workshop on FPGAs for Custom Computing Machines, Napa CA, April 10-13, pp. 189–198 (1994)
2. Bland, I.M., Megson, G.M.: Implementing a generic systolic array for genetic algorithms. In: Proc. 1st. On-Line Workshop on Soft Computing, pp. 268–273 (1996)
3. Brown, B.D., Card, H.C.: Stochastic Neural Computation I: Computational Elements. IEEE Transactions on Computers 50(9), 891–905 (2001)
4. Gaines, B.R.: Stochastic Computing Systems. Advances in Information Systems Science (2), 37–172 (1969)
5. Liu, J.: A general purpose hardware implementation of genetic algorithms, MSc. Thesis, University of North Carolina (1993)
6. MathWorks (2004), http://www.mathworks.com/
7. Megson, G.M., Bland, I.M.: Synthesis of a systolic array genetic algorithm. In: Proc. 12th. International Parallel Processing Symposium, pp. 316–320 (1998)
8. Michalewics, Z.: Genetic algorithms + data structures = evolution programs, 2nd edn. Springer, Berlin (1994)
9. Nedjah, N., Mourelle, L.M.: Reconfigurable Hardware Architecture for Compact and Efficient Stochastic Neuron. In: Mira, J., Álvarez, J.R. (eds.) IWANN 2003. LNCS, vol. 2687, pp. 17–24. Springer, Heidelberg (2003)
10. Scott, S.D., Samal, A., Seth, S.: HGA: a hardware-based genetic algorithm. In: Proc. ACM/SIGDA 3rd International Symposium in Field-Programmable Gate Array, pp. 53–59 (1995)
11. Scott, S.D., Seth, S., Samal, A.: A hardware engine for genetic algorithms. Technical Report, UNL-CSE-97-001, University of Nebraska-Lincoln (July 1997)
12. Turton, B.H., Arslan, T.: A parallel genetic VLSI architecture for combinatorial real-time applications – disc scheduling. In: Proc. IEE/IEEE International Conference on Genetic Algorithms in Engineering Systems, pp. 88–93 (1994)
13. Xilinx (2004), http://www.xilinx.com/

Chapter 2
Genetic Algorithms on Network-on-Chip*

Abstract. The aim of the work described in this chapter is to investigate migration strategies for the execution of parallel genetic algorithms in a Multi-Processor System-on-Chip (MPSoC). Some multimedia and Internet applications for wireless communications are using genetic algorithms and can benefit of the advantages provided by parallel processing on MPSoCs. In order to run such algorithms, we use a Network-on-Chip platform, which provides the interconnection network required for the communication between processors. Two migration strategies are employed, in order to analyze the speedup and efficiency each one can provide, considering the communication costs they require.

2.1 Introduction

The increasing demand of electronic systems, that require more and more processing power, low energy consumption, reduced area and low cost, has lead to the development of more complex embedded systems, also known as System-on-Chip (SoC), in order to run multimedia, Internet and wireless communication applications [9]. These systems can be built of several independent subsystems, that work in parallel and interchange data. When these systems have more than one processor, they are called Multi-Processor System-on-Chip (MPSoC).

Currently, several products, such as cell phones, portable computers, digital televisions and video games, are built using embedded systems. While in embedded systems the communication between Intellectual Property (IP) blocks is basically done through a shared bus, in multiprocessor embedded systems this kind of interconnection compromises the expected performance [2]. In this case, the communication is best implemented using an intrachip network, implemented by a Network-on-Chip (NoC) [6] [5] [1] platform.

Some multimedia and Internet applications for wireless communications are using genetic algorithms and can benefit from the advantages provided by parallel processing on MPSoCs. In this chapter, we present a parallel genetic algorithm that

---

* This chapter was developed in collaboration with Rubem Euzébio Ferreira.

N. Nedjah and L. de Macedo Mourelle, *Hardware for Soft Computing and Soft Computing for Hardware*, Studies in Computational Intelligence 529,
DOI: 10.1007/978-3-319-03110-1_2, © Springer International Publishing Switzerland 2014

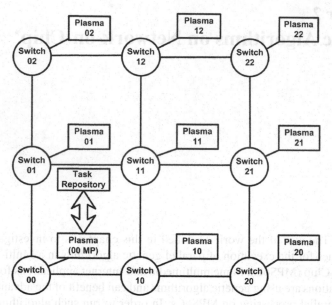

Fig. 2.1 HMPS architecture, with 9 RISC Plasma processors connected to a 3×3 mesh network

runs on Hermes Multi-Processor System (HMPS) architecture and discuss the impact of migration strategies on performance. In Section 2, we describe the HMPS architecture. The parallel genetic algorithm, used in this chapter, is presented in Section 3 and some simulation results are introduced in Section 4. Finally, we draw some conclusions and future work in Section 5.

2.2 Multi-processor System-on-Chip Platform

Figure 2.1 shows the Multi-Processor System-on-Chip (MPSoC), called Hermes Multiprocessor System (HMPS) [3]. MPSoC architectures may be represented as a set of processing nodes that communicate via a communication network. Switches compose the network and RISC processors the processing nodes (Plasma). Information exchanged between resources are transfered as messages, which can be split into smaller parts called packages [7]. The switch allows for retransmission of messages from one module to another and decides which path these messages should take. Each switch has a set of bidirectional ports for the interconnection with a resource and the neighboring switches.

As the total number of tasks composing the target application may exceed the MPSoC memory resources, one processor is dedicated to the management of the system resources (MP - Manager Processor). The MP has access to the task repository, from where tasks are allocated to some processors of the system.

The interconnection network is based on HERMES [4], that implements wormhole packet switching with a 2D-mesh topology. The HERMES switch employs input buffers, centralized control logic, an internal crossbar and five bi-directional ports. The Local port establishes the communication between the switch and its local IP core. The other ports of the switch are connected to neighboring switches. A centralized round-robin arbitration grants access to incoming packets and a deterministic XY routing algorithm is used to select the output port.

The processor is based on the PLASMA processor [10], a RISC microprocessor. It has a compact instruction set comparable to a MIPS-1, 3 pipeline stages, no cache, no Memory Management Unit (MMU) and no memory protection support in order to keep it as small as possible. A dedicated Direct Memory Access (DMA) unit is also used for speeding up task mapping, but not for data communications. The processor local memory (1024 Kbytes) is divided into four independent pages. Page 0 receives the microkernel and pages 1 to 3 the tasks. Each task can hold 256 Kbytes (0x40000).

The HMPS communication primitives, *WritePipe*() and *ReadPipe*(), essentially abstract communications, so that tasks can communicate with each other without knowing their position on the system, either on the same processor or a remote one. When HMPS starts, only the microkernel is loaded into the local memory. All tasks are stored in the task repository. The manager processor is responsible for reading the object codes from the task repository and transmit them to the other processors. The DMA module is responsible for transferring the object code from the network interfaces to the local memory.

2.3 Parallel Genetic Algorithm

The Parallel Genetic Algorithm (PGA) is based on the island model, in which serial isolated subpopulations evolve in parallel and each one is controlled by a single processor. This processor periodically sends its best individuals to neighboring subpopulations and receives their best individuals. These individuals are used to substitute the local worst ones. It is obvious that the GA time processing increases with population size. Therefore, small subpopulations tend to converge quickly when isolated.

The PGA is executed by the HMPS platform. Each processor corresponds to an island and its initial subpopulation is randomly generated, evolving independently from the other subpopulations, until the migration operator is activated, as described in Algorithm 2.1. Premature convergence occurs less in a multi-population GA and can be ignored, when other islands produce better results. Each island can use a different set of GA operators, i.e. crossover and mutation rates, which causes different convergence. Migration of the chromosomes among the islands prevents mono-race populations, which converge prematurely. Periodic migration, which occurs after some generations, prevents a common convergence among the islands.

Algorithm 2.1. PGA

  Initialize the evolutionary parameters
  $t \leftarrow 0$
  Initialize a random population $p(t)$
  Evaluate $p(t)$ in order to find th best solution
  while $(t < NumGenerations)$ **do**
    $t \leftarrow t + 1$
    Select $p(t)$ from $p(t-1)$
    Crossover
    Mutation
    Evaluate $p(t)$ in order to find th best solution
    if $(t \bmod MigrationRate = 0)$ **then**
      Migrate local $best[p(t)]$ to the next processor
      Receive remote $best[p(t)]$ from the previous processor
      Replace $worst[p(t)]$ by $best[p(t)]$
    end if
  end while

The PGA requires the definition of some parameters: number of processors, how often the migration will take place, which individuals will migrate and which individuals will be replaced due to migration. The island model introduces a migration operator in order to migrate the best individuals from one subpopulation to another.

2.3.1 Topology Strategies

In this work, we investigate two topology strategies to migrate individuals from one subpopulation to another: ring and neighborhood. In the ring topology, the best individuals from one subpopulation can only migrate to an adjacent one. As seen in Figure 2.2, the best individuals from subpopulation 6 can only migrate to subpopulation 1 and the best individuals from subpopulation 1 can only migrate to subpopulation 2. In Algorithm 2.2, migration is implemented using this kind of strategy. In the neighborhood topology, the best individuals from one subpopulation can migrate to a left and to a right neighbor, as seen in Figure 2.3. For this kind of strategy, migration is implemented as in Algorithm 2.3.

Choosing the right time of migration and which individuals should migrate are two critical decisions. Migrations should occur after a time long enough for allowing the development of good characteristics in each subpopulation. Migration is a trigger for evolutionary changes and should occur after a fixed number of generations in each subpopulation. The migrant individuals are usually selected from the best individuals in the origin subpopulation and they replace the worst ones in the destination subpopulation. Since there are no fixed rules that would give good results, intuition is still strongly recommended to fix the migration rate [11].

Sending an individual from one subpopulation to another increases the fitness of the destination subpopulation and maintains the population diversity of the other subpopulation. As in the sequential GA, issues of selection pressure and diversity

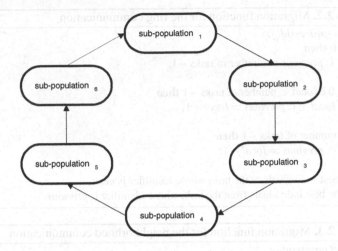

Fig. 2.2 Ring migration topology

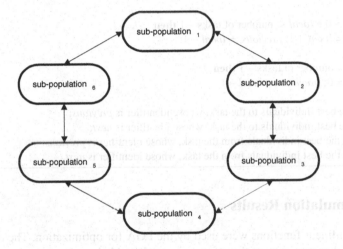

Fig. 2.3 Neighborhood migration topology

arise. If a subpopulation receives frequently and consistently highly fit individuals, these become predominant in the subpopulation and the GA will focus its search on them at the expense of diversity loose. On the other hand, if random individuals are received, the diversity may be maintained, but the fitness of the subpopulation may not be improved as desired. As migration policy, the best individual is chosen as the migrant, replacing the worst one in the receiving subpopulations. For the migration frequency, an empirical value was adopted based on the number of generations.

Algorithm 2.2. Migration function for the ring communication

$local := get\,processid();$
if $local = 0$ **then**
    $next := 1; previous :=$ number of tasks $-1;$
end if
if $local > 0$ e $local <$ number of tasks -1 **then**
    $next := local + 1; previous := local - 1;$
end if
if $local =$ number of tasks -1 **then**
    $next := 0; previous := local - 1;$
end if
Send the best individuals to the task, whose identifier is *next*;
Receive the best individuals from the task, whose identifier is *previous*.

Algorithm 2.3. Migration function for the neighborhood communication

$local := get\,processid();$
if $local = 0$ **then**
    $next := 1; previous :=$ number of tasks $-1;$
end if
if $local > 0$ e $local <$ number of tasks -1 **then**
    $next := local + 1; previous := local - 1;$
end if
if $local =$ number of tasks -1 **then**
    $next := 0; previous := local - 1;$
end if
Send the best individuals to the task, whose identifier is *previous*;
Send the best individuals to the task, whose identifier is *next*;
Receive the best individuals from the task, whose identifier is *previous*;
Receive the best individuals from the task, whose identifier is *next*.

2.4 Simulation Results

Three non-linear functions were used by the PGA for optimization. The definition and main characteristics of these functions are listed below

- Function $f_1(x)$ is defined in (2.1). This function plots into the curve depicted in Fig. 2.4. It presents 14 local maximum e one global maximum in the interval [-1, 2], with an approximate global maximum of 2.83917, at $x = 1.84705$.

$$\max_x f_1(x) = sen(10\pi x) + 1 \qquad (2.1)$$

- Function $f_2(x,y)$ is defined in (2.2). This function plots into the curve depicted in Fig. 2.5. It has many local minimum and one global minimum in the interval $-3 \leq x \leq 3$ and $-3 \leq y \leq 3$, and an approximate global minimum of -12.92393, at $x = 2,36470$ and $y = 2.48235$.

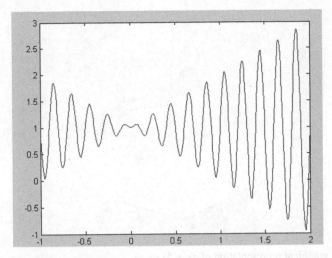

Fig. 2.4 The graphical representation of $f_1(x)$

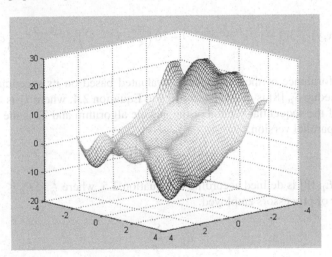

Fig. 2.5 The graphical representation of $f_2(x, y)$

$$\min_{x,y} f_2(x,y) = cos(4x) + 3sen(2y) + (y-2)^2 - (y+1) \qquad (2.2)$$

- Function $f_3(x, y)$ is defined in (2.3). This function plots into the curve depicted in Fig. 2.6. It has 2 local maximum and one global minimum in the interval $-3 \leq x \leq 3$ and $-3 \leq y \leq 3$, and an approximate global maximum of $8,11152$, at $x = 0,01176$ and $y = 1,58823$.

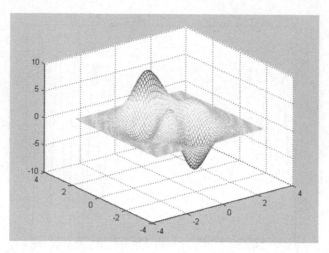

Fig. 2.6 The graphical representation of $f_3(x, 5)$

$$\max_{x,y} f_3(x,y) = 3(1-x)^2 e^{(-x^2-(y+1)^2)} - 10\left(\frac{x}{5} - x^3 - y^5\right) e^{(-x^2-y^2)} - \frac{1}{3} e^{-(x+1)^2-y^2}$$

$$(2.3)$$

The performance of the PGA can be evaluated based on its speedup and efficiency. Speedup S_p [8] is defined according to Equation 2.4, where T_1 is the execution time of the sequential version of the genetic algorithm and T_p is the execution time of its parallel version.

$$S_p = \frac{T_1}{T_p}$$

$$(2.4)$$

Efficiency E_p [8] is defined according to Equation 2.5, where $\frac{1}{p} < E_p \leq 1$ and p is the number of processors employed.

$$E_p = \frac{S_p}{p}$$

$$(2.5)$$

Table 2.1, Table 2.2 and Table 2.3 show the simulation results for the optimization of functions $f_1(x)$, $f_2(x,y)$ e $f_3(x,y)$ respectively using the ring topology for migration of individuals. In those tables, N_p is the number of used processors, M_r is the migration rate, M_i is the migration interval in terms of generation number, S_p is the speedup obtained and E_p is the efficiency yield for each used processor.

Table 2.4, Table 2.5 and Table 2.6 show the simulation results for the optimization of functions $f_1(x)$, $f_2(x,y)$ e $f_3(x,y)$ respectively using the neighborhood topology for migration of individuals. In those tables, N_p is the number of used processors, M_r is the migration rate, M_i is the migration interval in terms of generation number, S_p is the speedup obtained and E_p is the efficiency yield for each used processor.

Based on simulation results for the optimization of $f_1(x)$, $f_2(x,y)$ and $f_3(x,y)$ using the ring and neighborhood topologies, we obtained the graphics for speedup

Table 2.1 Simulation results for the optimization of function $f_1(x)$ for ring migration topology

N_p	M_r	M_i	Time (ms)	S_p	E_p
1	–	–	1127,5724	1	1
6	1	1	168,57284	6,68893	1,67223
		2	298,76094	3,77416	0,94354
	2	1	650,70556	1,73284	0,43321
		2	267,11808	4,22125	1,05531
9	1	1	112,10709	10,05799	1,25724
		2	102,16839	11,03641	1,37955
	2	1	381,15057	2,95834	0,36979
		2	101,16498	11,14587	1,39323
16	1	1	83,86655	13,44484	0,89632
		2	75,29244	14,97590	0,998393
	2	1	73,95938	15,24583	1,01638
		2	77,13687	14,61781	0,97452

Table 2.2 Simulation results for the optimization of function $f_2(x,y)$ for ring migration topology

N_p	M_r	M_i	Time (ms)	S_p	E_p
1	–	–	6024,11201	1	1
6	1	1	2569,06697	2,344863	0,586215
		2	2616,76305	2,302123	0,575530
	2	1	2507,48402	2,402452	0,600613
		2	1448,84485	4,157872	1,039468
9	1	1	1968,24989	3,06064	0,38258
		2	1250,55945	4,81713	0,60214
	2	1	1352,18413	4,45509	0,55688
		2	1112,49588	5,41495	0,67686
16	1	1	718,73197	8,38158	0,55877
		2	797,31202	7,55552	0,50370
	2	1	596,06991	10,10638	0,67375
		2	866,79268	6,94988	0,46332

and efficiency shown in Figure 2.7, Figure 2.8 and Figure 2.9 respectively. The data are presented as triples consisting of the number of slave processors used N_p, the migration rate M_r and the migration interval M_i.

2.5 Summary

For the ring topology, the behavior of the two functions shows that, keeping the migration interval constant and varying the migration rate, if the increase in the migration rate resulted in an increase in speedup and efficiency, the fitness of

Table 2.3 Simulation results for the optimization of function $f_3(x, y)$ for ring migration topology

N_p	M_r	M_i	Time (ms)	S_p	E_p
1	–	–	6209,50022	1	1
6	1	1	2778,47764	2,23485	0,55871
		2	2927,64913	2,12098	0,53024
	2	1	3143,09053	1,97560	0,49390
		2	2925,58322	2,12248	0,53062
9	1	1	1037,66721	5,98409	0,74801
		2	1832,88554	3,38782	0,42347
	2	1	1799,06522	3,45151	0,43143
		2	1433,94829	4,33035	0,54129
16	1	1	873,31097	7,11029	0,47401
		2	723,58761	8,58154	0,57210
	2	1	607,38299	10,22336	0,68155
		2	942,71555	6,58682	0,43912

Table 2.4 Simulation results for the optimization of function $f_1(x)$ for neighborhood migration topology

N_p	M_r	M_i	Time (ms)	S_p	E_p
1	–	–	1127,5724	1	1
6	1	1	645,36593	1,74718	0,43679
		2	535,14461	2,10704	0,52676
	2	1	172,29855	6,54429	1,63607
		2	172,80265	6,52520	1,63130
9	1	1	217,88489	5,17508	0,64688
		2	304,68098	3,70082	0,46260
	2	1	104,90308	10,74870	1,34358
		2	188,31056	5,98783	0,74847
16	1	1	80,62822	13,98483	0,93232
		2	121,45834	9,28361	0,61890
	2	1	71,09218	15,86070	1,05738
		2	131,73707	8,55926	0,57061

the individuals, received by one or more populations during the migration phase, accelerated the evolutionary process, decreasing the convergence time. On the other hand, if the increase in the migration rate resulted in the decrease of speedup and efficiency, then we can say that the fitness of these individuals did not influence enough the evolutionary process of the populations that received them. In this case, the convergence time increases.

In the future, we intend to investigate the impact of other migration strategies on the performance of the parallel Network-on-chip based implementation of genetic algorithms. One of the these topologies is broadcasting, which allows each processor

Table 2.5 Simulation results for the optimization of function $f_2(x,y)$ for neighborhood migration topology

N_p	M_r	M_i	Time (ms)	S_p	E_p
1	–	–	6024,11201	1	1
6	1	1	2970,49815	2,02798	0,50699
		2	2241,80203	2,68717	0,67179
	2	1	2977,43556	2,02325	0,50581
		2	2635,64829	2,28562	0,57140
9	1	1	1560,87682	3,85944	0,48243
		2	1370,53135	4,39545	0,54943
	2	1	1772,67139	3,39832	0,42479
		2	1161,60725	5,18601	0,64825
16	1	1	719,73603	8,36989	0,55799
		2	951,55986	6,33077	0,42205
	2	1	574,84260	10,47958	0,69863
		2	700,59551	8,59855	0,57323

Table 2.6 Simulation results for the optimization of function $f_3(x,y)$ for neighborhood migration topology

N_p	M_r	M_i	Time (ms)	S_p	E_p
1	–	–	6209,50022	1	1
6	1	1	2534,68066	2,44981	0,61245
		2	2497,41481	2,48637	0,62159
	2	1	3075,95908	2,01872	0,50468
		2	2737,40887	2,26838	0,56709
9	1	1	1698,95341	3,65489	0,45686
		2	1398,89571	4,43885	0,55485
	2	1	830,546335	7,47640	0,93455
		2	1296,38967	4,78984	0,59873
16	1	1	1235,58877	5,02553	0,33503
		2	910,60102	6,81912	0,45460
	2	1	777,34866	7,98805	0,53253
		2	683,45716	9,08542	0,60569

to send the best solution found so far to all the other processors in the network. We will assess the impact of heavy message send/receive workload on the overall system performance.

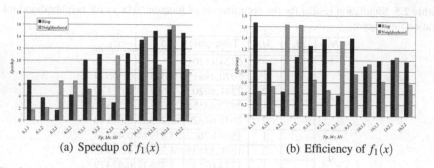

(a) Speedup of $f_1(x)$ (b) Efficiency of $f_1(x)$

Fig. 2.7 Impact of the migration rate and migration interval on speedup and efficiency for function $f_1(x)$, considering the used topology

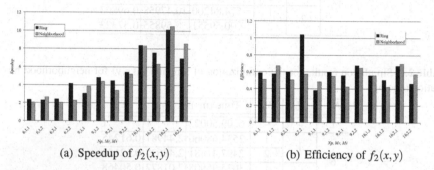

(a) Speedup of $f_2(x,y)$ (b) Efficiency of $f_2(x,y)$

Fig. 2.8 Impact of the migration rate and migration interval on speedup and efficiency for function $f_2(x,y)$, considering the used topology

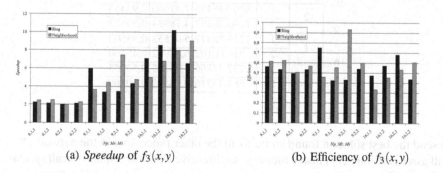

(a) *Speedup* of $f_3(x,y)$ (b) Efficiency of $f_3(x,y)$

Fig. 2.9 Impact of the migration rate and migration interval on speedup and efficiency for function $f_3(x,y)$, considering the used topology

References

1. Ivanov, A., De Micheli, G.: The network-on-chip paradigm in practice and research. IEEE Design and Test of Computers 1(1), 399–403 (2005)
2. Mello, A.M.: Arquitetura multiprocessada em SoCs: estudo de diferentes topologias de conexão (June 2003) (in Portuguese)
3. Woszezenki, C.: Alocação de tarefas e comunicação entre tarefas em mpsocs. M.Sc., Faculdade de Informática, PUCRS, Porto Alegre, RS, Brazil (June 2007) (in Portuguese)
4. Moraes, F., Calazans, N., Mello, A., Möller, L., Ost, L.: Hermes: an infrastructure for low area overhead packet-switching networks on chip. Integration, The VLSI Journal 38(1), 69–93 (2004)
5. Öberg, J., Jantsch, A., Tenhunen, H.: Special issue on networks on chip. Journal of Systems Architecture 1(1), 61–63 (2004)
6. Beniniand, L., De Micheli, G.: Networks on chips: a new soc paradigm. IEEE Computer 1(1), 70–78 (2002)
7. Benini, L., Ye, T.T., De Micheli, G.: Packetized on-chip interconnect communication analysis for MPSoC. In: Proceedings of the Design, Automation and Test in Europe Conference and Exhibition (DATE 2003), pp. 344–349. IEEE Press (2003)
8. Chiwiacowsky, L.D., de Campos Velho, H.F., Preto, A.J., Stephany, S.: Identifying initial conduction in heat conduction transfer by a genetic algorithm: a parallel approach 28, 180–195 (1980)
9. Ruiz, P.M., Antonio: Using genetic algorithms to optimize the behavior of adaptive multimedia applications in wireless and mobile scenarios. In: IEEE Wireless Communications and Networking Conference (WCNC 2003), pp. 2064–2068. IEEE Press (2003)
10. Rhoads, S.: Plasma microprocessor (2009), http://www.opencores.org
11. Hue, X.: Genetic algorithms for optimization – background and applications. Technical report. Edinburgh Parallel Computer Centre, The University of Edinburgh (1997)

Chapter 3
A Reconfigurable Hardware for Particle Swarm Optimization*

Abstract. The Particle Swarm Optimization or PSO is a heuristic based on a population of individuals, in which the candidates for a solution of the problem at hand evolve through a simulation process of a social adaptation simplified model. Combining robustness, efficiency and simplicity, PSO has gained great popularity as many successful applications are reported. The algorithm has proven to have several advantages over other algorithms that based on swarm intelligence principles. The use of PSO solving problems that involve computationally demanding functions often results in low performance. In order to accelerate the process, one can proceed with the parallelization of the algorithm and/or mapping it directly onto hardware. This chapter presents a novel massively parallel co-processor for PSO implemented using reconfigurable hardware. The implementation results show that the proposed architecture is very promising as it achieved superior performance in terms of execution time when compared to the direct software execution of the algorithm.

3.1 Introduction

Swarm Intelligence is a field of artificial intelligence wherein a decentralized collective behavior of individuals that interact with each other as well as with the environment is at the basis to infer an intelligent decision with respect to a given problem.

This is an innovative paradigm of distributed intelligence, inspired originally from biological groups such as flocks, herds and schools of fish. There are several models based on this concept, some describe swarm particles, others are based on groups and social behavior of humans in general. In addition, there are also models based on the social behavior of bacteria, spiders, bees and sharks, among others.

PSO was introduced by Kennedy and Eberhart [1] and is based on collective behavior, social influence and learning. PSO seeks to imitate the behavior of social groups of animals, specifically flocks of birds. If one element of the group discovers a way to get to get to the food source, the other group members tend instantly, to follow the indicated way. Many successful applications of PSO are

---

* This chapter was developed in collaboration with Rogério de Moraes Calazan.

N. Nedjah and L. de Macedo Mourelle, *Hardware for Soft Computing and Soft Computing for Hardware*, Studies in Computational Intelligence 529,

DOI: 10.1007/978-3-319-03110-1_3, © Springer International Publishing Switzerland 2014

reported, in which this algorithm has shown several advantages over other algorithms that are based on swarm intelligence. It is robust, efficient and simple. Moreover, it usually requires less computational effort when compared to other evolutionary algorithms [2].

FPGAs (Field Programmable Gate Arrays) represent a class of integrated circuits designed to be user-configured after manufacture. FPGAs present several advantages over other alternatives for hardware implementation, such as reduced time-to-market, use the synthesis of very flexible specification, allowing a very low development cost. Programming these devices is done through hardware description languages (HDL), such as Verilog [18] and VHDL (Very High Speed Integrated Circuits Hardware Description Language) [19]. The first commercial FPGA (XC2064 model) was developed by XILINX and marketed in 1985. Since then the number of gates followed an exponential growth rate. These reconfigurable devices are suitable for applications that exploit the characteristics of the inherent massively parallel structures by relocate computational-intensive parts of the implementation into the FPGA and thus allowing parallel processing. In general, parallel computation in FPGA allow for a considerable throughput performance at low clock rates, which in turn occasion very low power consumption. With adequate development environments, the task of compiling a hardware specification, synthesis and downloading the result to FPGA chip became relatively simple.

With the continuous advancement in FPGA technology, providing increased performance and high-density devices, processor can be offered as an economic alternative. The $MicroBlaze^{TM}$ embedded processor [15] has a reduced instruction set (RISC) and is optimized for Xilinx®FPGAs implementations.

The purpose of this chapter is to present an efficient hardware architecture for parallel PSO. The proposed implementation uses floating-point arithmetics. The architecture is viewed as a co-processor that operates together with the MicroBlaze processor in order to solve specific applications, optimize the performance and free up the processor during execution of PSO steps. In order to evaluate the performance of the proposed architecture, we compare the execution time of PSO with and without the co-processor, and we test the impact of dynamic update of the inertia weight parameter of the PSO algorithm.

This chapter is organized as follows: First, in Section 3.2, we present some related works; Then, in Section 3.3, we sketch briefly the PSO process and the algorithm; In Section 3.4 presents the architecture of the MicroBlaze embedded processor and Thereafter, in Section 3.5, we describe the implementation of the proposed co-processor and its architecture; Subsequently, in Section 3.6, we report and evaluate the performance of the co-processor; Finally, in Section 8.5, we draw some concluding remarks and point out directions for future work.

3.2 Related Works

In [5], the PSO algorithm together with some test functions are implemented in an FPGA using Floating-point operations. The performance results show that the proposed implementation can be up to 78 times faster than a MATLAB software implementation. In[10] and [11], the authors present a Parallel architecture for the PSO algorithm. The implementation was done using the High-Performance MPI (Message Passing Interface). Both synchronous and asynchronous solutions were investigated. The proposed implementation showed an improvement in processing time for a bio-mechanical test problem. An hardware/software co-design architecture to implement the PSO algorithm in an FPGA is reported in [7]. The particle accelerator module was implemented in hardware while the fitness function was kept in software. Whenever complex fitness functions, which impose high computational costs, this implementation showed poor performance as the critical point is the fitness function and not in the rest of the PSO algorithm. An FPGA implementation of Simultaneous Perturbation PSO (SPPSO) is presented in [9]. The authors were able to increase the operating speed using parallelism within the PSO process. In [4], a performance comparison of the PSO algorithm implemented using a 16-bit microcontroller (Freescale MC9512DP256) and an direct implementation in hardware is described.

3.3 Particle Swarm Optimization

The main steps of the PSO algorithm are desribed in Algorithm 3.1. Note that, in this specification, the computations are executed sequentially. In this algorithm, each particle has a *velocity* and an ıadaptive direction [1] that determines its next movement within the search space. The particle is also endowed with a memory that makes it able to remember the best previous position that it passed by.

Algorithm 3.1. PSO

---
    for $i = 1$ *to n_particles* **do**
        Initialize the information of particle i
        Randomly initialize position and velocity of particle i;
    end for;
    repeat
        for $i = 1$ *to n_particles* **do**
            Compute the *Fitness$_i$* of particle i;
            if *Fitness$_i$* $\leq$ *Pbest* **then**
                Update *Pbest* using the position of particle i;
            end if;
            if *Fitness$_i$* $\leq$ *Gbest* **then**
                Update *Gbest* using the position of particle i;
            end if;
            Update the velocity of particle i;
            Update the position of particle i;
        end for;
    until Stopping condition is true
    return *Gbest* and corresponding position;

The PSO is formed by a set of particles, each of which represents a potential solution to the problem, having position coordinates in a space of n-dimensional search. Thus, each particle is represented by a current position vector, a vector of best position found by the particle so far, one field to store the fitness and another for best fitness. To update the position of each particle i, there is a set of velocities, one for each dimension j of this position. The velocity is the element that promotes the ability of movement of the particles, and can be calculated according to Equation 3.1 while the position is defined as in Equation 3.2.

3.3.1 Global Best PSO

In this variation of the PSO algorithm, the neighborhood of each particle is formed by all the population particles. Thus, it can be viewed as the star topology, as shown in Figure 3.1. Using this strategy, the social component of the particle velocity is influenced by all other particles [2] [8]. The velocity is the element that promotes the capacity of particle locomotion and can be computed as described in (3.1) [1] [2], wherein w is called *inertia weight*, r_1 and r_2 are random numbers in [0,1], c_1 and c_2 are positive constants, y_{ij} is the best position *Pbest* found by the particle i so far w.r.t. dimension j and y_j is the best position *Gbest* w.r.t. dimension j, found so far, considering all the population particles. The position of each particle is updated according as described in (3.2). Note that $x_{ij}(t+1)$ is the current position and $x_{ij}(t)$ is the previous position.

$$v_{ij}(t+1) = wv_{ij}(t) + c_1 r_1 \left(y_{ij} - x_{ij}(t)\right) + c_2 r_2 \left(y_j - x_{ij}(t)\right) \qquad (3.1)$$

$$x_{ij}(t+1) = v_{ij}(t+1) + x_{ij}(t) \qquad (3.2)$$

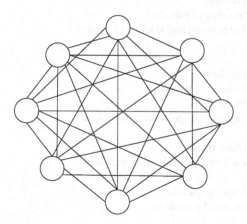

Fig. 3.1 Star Social Structure

The velocity component drives the optimization process, reflecting both the experience of the particle, and the exchange of information between the particles. The particle experimental knowledge is referred to the cognitive behavior, which is proportional to the distance between the particle and its best position found w.r.t. its first iteration [8]. The exchange of information between particles refers to the social behavior of the velocity definition. After upgrading the velocity and position of each particle it checks the stopping criteria and then displays the result or perform another operation.

The value of each parameter of the PSO algorithm is crucial in the search process, and therefore the importance of defining appropriate values at the initialization step. The inertia weight w was introduced by [3] as a mechanism to control the exploration and exploitation abilities of the swarm. Large values for w facilitate exploration, with increased diversity. A small values for w promotes local exploitation [2]. Values greater than 1 tend to leave the particles with a very high acceleration, while lower values, approximately 0, can decelerate too much the search. Dynamic update of w has also been used to adapt the search velocity. Starting with a high value which gradually decreases during the optimization process.

The cognitive coefficient c_1 and the coefficient social c_2 cause the algorithm to perform better if they are balanced, i.e. $c_1 = c_2$ [2]. Also according to [2], recent studies indicate that we should have $c_1 + c_2 = 4$, and that good results were achieved using $c_1 = c_2 = 1.49$. The factors r_1 and r_2 define the stochastic content of cognitive and social contributions of the algorithm. Random values are selected in the range [0,1] [2] for each of the factors.

The maximum velocity is defined for each dimension of the search space, and can be formulated as a percentage of the domain, according to Equation 3.3, where x_{max} and $x_{m}in$ are respectively the maximum and minimum value of the domain and δ is a value in the interval [0,1].

$$v_{max} = \delta(x_{max} - x_{min}) \tag{3.3}$$

The amount of particles defines the possibility of covering a certain portion of the search space in each iteration. A large number of particles allows for greater coverage, but requires more computing effort. According to [2], empirical studies have shown that the PSO can reach optimal solutions using a small number of particles, between 10 and 30.

3.3.2 Parallel PSO

Aiming at improving the performance of the PSO, Algorithm 3.1 has been parallelized as shown in Figure 3.2. Each particle performs its fitness velocity and position calculations, independently and in parallel with the other particles until the election of *Gbest*. In order to synchronize the process and prevent using incorrect values *Gbest*, the velocity and position computations can only commence once *Gbest* have been chosen among *Pbest* of all particles.

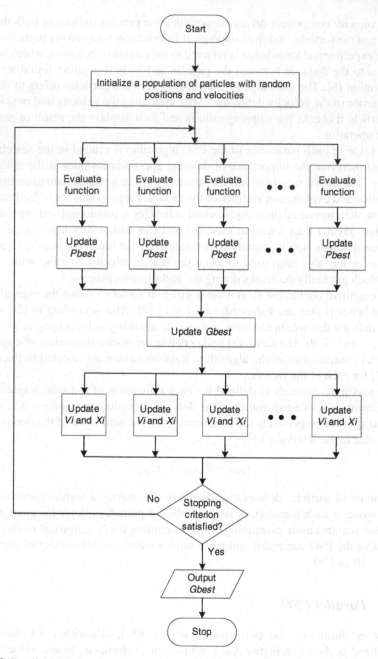

Fig. 3.2 Parallel PSO data flow

3.4 The MicroBlaze Embedded Processor

FPGAs are programmable logic devices, in which a matrix of Configurable Logic Blocks or CLBs, connected by programmable channels can be configured to perform a set of logical functions and give rise to any type of digital system. They are excellent means of prototyping [14].

The advantages of using an FPGA to implement a digital system goes beyond the benefits of prototyping. Depending on the amount of resources (logical blocks) available on a chip, you can replicate various components of the system architecture and perform operations in parallel.

The *MicroBlaze*^{TM} embedded processor soft core is a reduced instruction set computer (RISC) optimized for implementation for *Xilinx*^{TM} FPGAs. The fixed feature set of the processor includes Thirty-two 32-bit general purpose registers, 32-bit instruction word with three operands and two addressing modes, 32-bit address bus, single issue pipeline and other features [15]. The Figure 3.3 shows a functional block diagram of the MicroBlaze core.

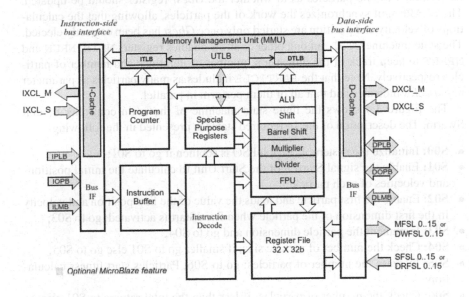

Fig. 3.3 MicroBlaze Core Block Diagram

The result of the computation can be transmitted to a workstation via a UART (Universal Assynchronous Receiver/Transmission) interface [17] and then, interpreted by the embedded software.

The XPS UART Lite is an example of component of *Xilinx*^{TM} intellectual property that connects to the PLB (Processor Local Bus) bus of MicroBlaze and performs this communication interfacing between the processor and the workstation.

The programming of the microprocessor can be done in ANSI C++ and compiled by XPS (Xilinx Studio Plataform), which generates for each component included the respective control data.

3.5   Co-processor Architecture

The main component of the co-processor architecture is depicted in Figure 3.5, and termed the SWARM unit. It is responsible for the correct operation and synchronization of the swarm. It starts by enabling the START unit that generates the initial position and velocity of each particle in the population. The stochastic nature of the PSO algorithm requires the use of random numbers generator. An LFSR (Linear Feedback Shift Register) is responsible in generating random numbers for single precision. This is done according to the maximum and minimum value of the domain as well as on number of particles allowed. As these values are loaded, the unit enables the particles to start the fitness calculation. Whenever the particle is ready to informs the value of the respective *Pbest*, the comparator checks, the values returned by the particles as to whether the *Gbest* register should be updated. The SWARM unit synchronizes the work of the particles, allowing that the calculations of velocity and position are started only once *Gbest* has been correctly elected. The state machine CTRL, among other controls enables registers NDIM, NRUN and NPART to keep track of the dimension, number of iterations and number of particles respectively. Note that the PARTICLE includes as many particles as parameter *n_particles* indicates and that all of them perform in parallel.

The Figure 3.4 shows the finite state machine of the main controller of unit Swarm. The description of the 14 required states is presented in the following:

- **S00:** Initialize the system; If EnablePSO is set then it go to S01;
- **S01:** Enable the signal StartEn of the Start Unit to calculate the initial positions and velocities of each particle;
- **S02:** Enable the first particle and loads the value of the first position and velocity in the first dimension of the particle when ReadStart is activated; go to S03;
- **S03:** Increment the particle dimension and go to S04;
- **S04:** Check the number of particle size, if smaller go to S01 else go to S05;
- **S05:** Increase the number of particles; go to S06; Particles start fitness calculation;
- **S06:** Check the number of particles. If less than the total returns to S01 else go to S07;
- **S07:** Initialize the particle counter; go to S08;
- **S08:** Wait in S08 until the signal ReadyPart of the first particle is active (the result of the particle fitness is ready); go to S09;
- **S09:** Compare the Fitness value of the particle with the value of Global Fitness; if smaller write the new fitness values and the coordinates in registers Gbest1 and Gbest2; go to S10;
- **S10:** Increment counter of particles; go to S11;
- **S11:** Check the number of particles; If less return to S09 else go to S12;

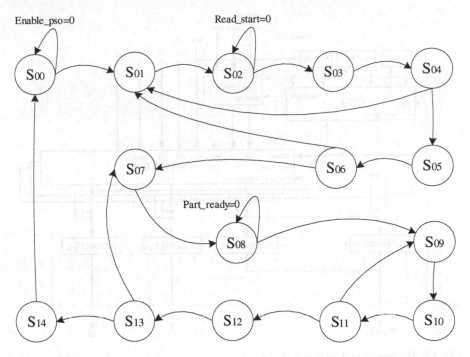

Fig. 3.4 Swarm FSM

- **S12:** Enable signal "readgbest"; particles update velocity (Equation 3.1) and position (Equation 3.2);
- **S13:** Check the number of iterations, if less return to S07 else go to S14;
- **S14:** Display the final result; go to S00.

The PARTICLE unit of Figure 3.6 is equipped with an internal memory represented by a bank of registers: two for each dimension (velocity and position), an execution module PSO CORE for computing the velocity and position of the particle, a module for calculating the fitness function and a single-precision floating-point unit FPU (IEEE 754 standard) to actually perform the calculations. Note that there are as many PARTICLE units as particles in the swarm, and thus enabling the massively parallel execution of particle-related operations. After loading the initial values and triggering, the FITNESS unit commences the calculation of the fitness function value. The result of this operation is then compared with the initial value and updated whenever necessary. At this point, the particle *Pbest* value is passed to the SWARM unit that, after comparing it with the results of other particles, elects the adequate *Gbest*, passes the new value through to all existing PARTICLE units, hence triggering the respective PSO CORE modules. The latter performs the computation of the new particle velocity and position in the search space. This done for

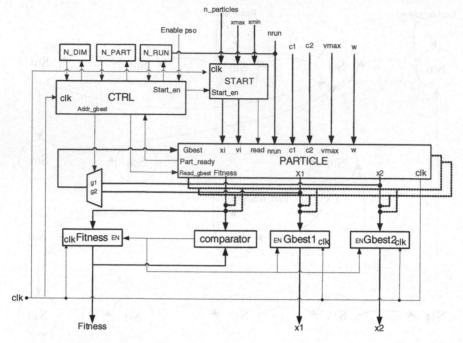

Fig. 3.5 The Swarm unit architecture

each dimension and according to (3.1) and (3.2). At this stage, the particle is ready
to either perform a new iteration or terminate the optimization process according to
the stopping criterion adopted.

3.6 Performance Results

The *MicroBlaze*^{TM} and the co-processor PSO were synthesized in a Xilinx Virtex
5 FPGA xc5vfx70t. The MicroBlaze embedded processor soft core is a reduced
instruction set computer (RISC) optimized for implementation in $Xilinx^{TM}$ FPGAs.
    Without the proposed PSO co-processor, the MicroBlaze processor performs
all computing. In this case, the PSO algorithm together with the fitness functions used were implemented in ANSI/C++. The MicroBlaze has a communication
interface for point-to-point, called Fast Simplex Link (FSL), which allows for an efficient connection with an external component, termed as the co-processor. The co-
processor, besides the fact that is being used as a hardware accelerator, it increases
the CPU availablability to perform other critical activities while the co-processor
performs the PSO specific computation. Figure 3.7 shows the processor as it is connected to the PSO co-processor hardware.

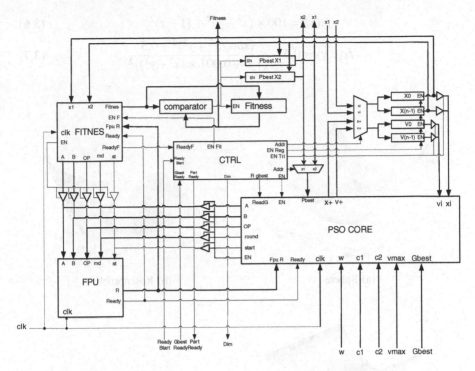

Fig. 3.6 Particle unit architecture

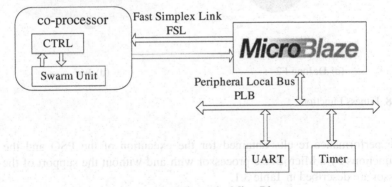

Fig. 3.7 PSO co-processor connected by FSL to the MicroBlaze

Four functions are implemented in order to assess the perfromance of the proposed PSO co-processor architecture: The Sphere as defined in (3.4), Rosenbrock Function (3.5), DeJong F2 Function (3.6) and F6 Function (3.7).

$$f_1(x,y) = f(x,y) = x^2 + y^2 \qquad (3.4)$$

$$f_2(x,y) = 100 \times (y - x^2)^2 + (x - 1)^2 \qquad (3.5)$$

$$f_3(x) = 100 \times (x^2 - y)^2 + (1 - x)^2 \tag{3.6}$$

$$f_4(x,y) = 0.5 - \frac{(sin\sqrt{x^2+y^2})^2 - 0.5}{(1.0 + 0.001 \times (x^2+y^2))^2} \tag{3.7}$$

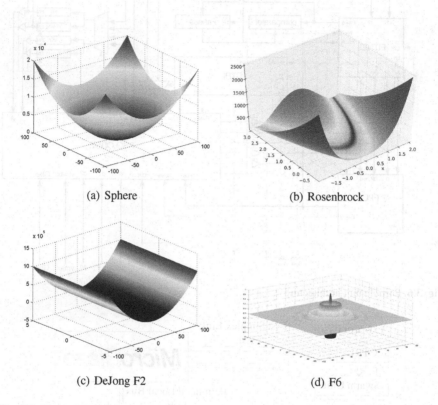

(a) Sphere   (b) Rosenbrock

(c) DeJong F2   (d) F6

Fig. 3.8 Fitness Function

The performance results obtained for the execution of the PSO and the fitness functions by a MicroBlaze processor with and without the support of the co-processor are described in Table 3.1.

Table 3.2 presents the synthesis results for the MicroBlaze processor using a swarm with 4 particles in a 32-bit architecture to solve the optimization problems of two dimensions. These results are viable for evaluating the efficiency of the architecture, the performance of circuits implemented, in addition to showing the resource requirements in the FPGA. Table 3.3 shows the comparison of the execution time for the conditions described in Table 3.1, as well as the performance obtained by the co-processor compared to MicroBlaze processor. In this table, one can note that the performance of PSO co-processor, specifically on a Xilinx Virtex 5 FPGA (xc5vfx70t) operating at 50 MHz, is up to 20 times faster than the implementation on MicroBlaze, which represents a very nice improvement indeed.

Table 3.1 Results for MicroBlaze processor *vs.* the PSO co-processor

Function	Range	Iterations	Co-processor (milliseconds)	MicroBlaze (milliseconds)
Sphere	[−100 100]	200	3.15	30.48
DeJong	[−5 5]	200	3.48	30.58
Rosenbrock	[−16 16]	500	9.82	202.82
F6	[−100 100]	500	76.77	343.56

Table 3.2 Synthesis results for four particles in two dimensions optimization problems

Implementation	Registers Max 44800	%	LuTs Max 44800	%	Frequency (MHz)
MicroBlaze*	2262	5	2627	5	100.00
Sphere[†]	15100	28	38498	85	86.55
DeJong[†]	15572	28	38670	86	93.08
Rosenbrock[†]	15168	29	38873	86	93.11
F6[†]	15168	33	38873	86	82.45

*Area requirements for the MicroBlaze only.
[†]Area requirements for the MicroBlaze and the PSO co-processor.

Table 3.3 Co-processor performance

Fitness Function	MicroBlaze (millisecond)	Co-processor (millisecond)	Performance Factor
Sphere	30.48	3.15	9.67
DeJong	30.58	3.48	8.78
Rosenbrock	202.82	9.82	20.65
F6	343.56	76.77	4.47

3.7 Summary

This chapter presents a parallel architecture of the PSO algorithm implemented as a hardware co-processor to the MicroBlaze processor. The FPGA used is a Xilinx Virtex 5 FPGA (xc5vfx70t). The architecture exploits the parallelism of updating the particle positions and velocities and getting the result of the fitness function independently of the others particles of the swarm. The synthesis results show that the scalability of the hardware depends on the number of particles used and the complexity of the fitness function. The architecture of PSO was validated using four particles operating in parallel to solve optimization problems of two dimensions. The best acceleration was obtained for the Rosenbrock function, achieving a performance of 20 times faster.

Further investigation of the impact of the co-processor will be carried out. For instance, we intend to check out the impact of the use of several PSO co-processors, each one searching in one voxel of a many-voxels search space.

References

1. Kennedy, J., Eberhart, R.: Particle Swarm Optimization. In: IEEE International Conference on Neural Network (1995)
2. Engelbrecht, A.P.: Computational Swarm Intelligence. In: Wiley Fundamentals of Computational Swarm Intelligence (2005)
3. Shi, Y., Eberhart, R.C.: A Modified Particle Swarm Optimizer. In: Proceedings of the IEEE Congress on Evolutionary Computation, pp. 69–73 (1998)
4. Tewolde, G.S., Hanna, D.M., Haskell, R.E.: Accelerating the Performance of Particle Swarm Optimization for Embedded Applications. In: Congress on Evolutionary Computation (2009)
5. Muoz, D.M., Llanos, C.H., dos Santos Coelho, L., Ayala-Rincn, M.: Hardware Architecture for Particle Swarm Optimization using Floating-Point Aritmetic. In: Ninth International Conference on Intelligent Systems Design and Applications, pp. 243–248 (2009)
6. Sadhasivam, G.S., Meenakshi, D.K.: Load Balance, Efficient Scheduling Whit Parallel Job Submission in Computational Grids Using Parallel Particle Swarm Optimization. In: World Congress on Nature e Biologically Inspired Computing, pp. 175–180 (2009)
7. Li, S.-A., Wong, C.-C., Yu, C.-J., Hsu, C.-C.: Hardware/Software Co-design for Particle Swarm. Jornal the National Science, 3762–3767 (2010)
8. Nedjah, N., dos Santos Coelho, L., de Macedo Mourelle, L. (eds.): Multi-Objective Swarm Intelligent Systems. SCI, vol. 261. Springer, Heidelberg (2010)
9. Maeda, Y., Matsushita, N.: Simultaneous Pertubation Particle Swarm Optimization Using FPGA. In: International Joint Conference on Neural Networks (August 2007)
10. Shutte, J.F., Reinbolt, J.A., Fregly, B.J., Haftka, R.T., George, A.D.: Parallel global optimization with the particle swarm algorithm. NIH Public Acsess, Int. J. Numer. Methods Eng., 2296–2315 (December 2004)
11. B.-l. Koh, A.D., George, R.T., Haftka, B.J.: Fregly: Parallel asynchronous particle swarm algorithm. NIH Public Acsess, Int. J. Numer. Methods Eng., 578–595 (July 2006)
12. Rosenbrock, H.H.: An automatic method for finding the greatest or least value of a function. The Computer Journal (1960)
13. Al-Eryani, J.: Floating Point Unit (2006)
14. XILINX Virtex-5 User Guide, Embedded Development Kit EDK 10.1i (2011),
    http://www.xilinx.com/support/documentation/
    user_guides/ug190.pdf
15. XILINX MicroBlaze Processor Reference Guide, v5.3 (2011),
    http://www.xilinx.com/support/documentation/
    sw_manuals/mb_ref_guide.pdf
16. XILINX Fast Simplex Link v2.11c (2011), http://www.xilinx.com/support/
    documentation/ip_documentation/fsl_v20.pdf
17. XILINX XPS UART Lite v1.01a (2011), http://www.xilinx.com/support/
    documentation/ip_documentation/xps_uartlite.pdf
18. Verilog Resources, Verilog Hardware Description Language (2011),
    http://www.verilog.com
19. EDA Industry Working Groups, VHDL – Very High Speed Integrated Circuits Hardware Description Language (2011), http://www.vhdl.org/

Chapter 4
Particle Swarm Optimization on Crossbar Based MPSoC*

Abstract. Multi-Processor System on Chip (MPSoC) offers a set of processors, embedded in one single chip. A parallel application can, then, be scheduled to each processor, in order to accelerate its execution. One problem in MPSoCs is the communication between processors, necessary to run the application. The shared memory provides the means to exchange data. In order to allow for non-blocking parallelism, we based the interconnection network in the crossbar topology. In this kind of interconnection, processors have full access to their own memory module simultaneously. On the other hand, processors can address the whole memory. One processor accesses the memory module of another processor only when it needs to retrieve data generated by the latter. This chapter presents the specification and modeling of an interconnection network based on crossbar topology. The aim of this work is to investigate the performance characteristics of a parallel application running on this platform.

4.1 Introduction

During the 80's and 90's, engineers were trying to improve the processing capability of microprocessors by increasing clock frequency [11]. Afterwards, they tried to explore parallelism at the instruction level with the concept of pipeline [1] [3]. However, the speedup required by software applications was gradually becoming higher than the speedup provided by these techniques. Besides this, the increase in clock frequency was leading to the increase in power required, to levels not acceptable. The search for smaller devices with high processing capability and with less energy consumption have turned solutions based on only one processor obsolete. This kind of solution has been restricted to low performance applications. On the other hand, there are few applications of this sort, for which microcontrollers are best employed.

In order to reach specific performance requirements, such as throughput, latency, energy consumed, power dissipated, silicon area, design complexity, response time, scalability, the concept of Multi-Processor System on Chip (MPSoC) was explored.

---

* This chapter was developed in collaboration with Fábio Gonçalves Pessanha.

N. Nedjah and L. de Macedo Mourelle, *Hardware for Soft Computing and Soft Computing for Hardware*, Studies in Computational Intelligence 529,
DOI: 10.1007/978-3-319-03110-1_4, © Springer International Publishing Switzerland 2014

In this concept, several processors are implemented in only one chip to provide the most of parallelism possible. MPSoCs require an interconnection network [7] to connect the processors, as shown in Fig. 4.1. Interconnection networks, beyond the context of MPSoCs, are implemented in different topologies, such as shared-medium, direct, indirect and hybrid [6] [4].

Another example is the crossbar, which connects processor-memory nodes through dedicated communication links, using a switch at each node [4], as depicted in Fig. 4.2 for a 4x4 network.

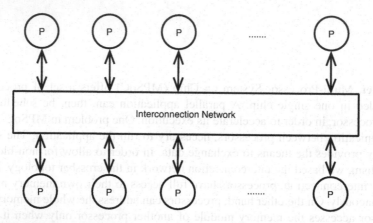

Fig. 4.1 Interconnection Network in a Multi-Processor System

4.2 The Crossbar Topology

The crossbar network allows for any processor to access any memory module simultaneously, as far as the memory module is free. Arbitration is required when at least two processors attempt to access the same memory module. However, contention is not an usual case, happening only when processors share the same memory resource, for example, in order to exchange information. In this work, we consider a distributed arbitration control, shared among the switches connected to the same memory module. In Fig. 4.3, the main components are introduced, labeled according to their relative position in the network, where i identifies the row and j identifies the column. For instance, component A(j) corresponds to the arbiter [5] for column j. For the sake of legibility, we consider 4 processors ($0 \leq i \leq 3$) and 4 memory modules ($0 \leq j \leq 3$).

The processor is based on the PLASMA CPU core, designated MLite_CPU (MIPS Lite Central Processor Unit) [8], shown in Fig. 4.4. In order to access a memory module M(j), processor P(i) must request the corresponding bus B(j) and

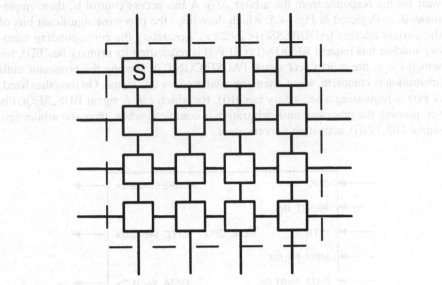

Fig. 4.2 Crossbar Switches Interconnection

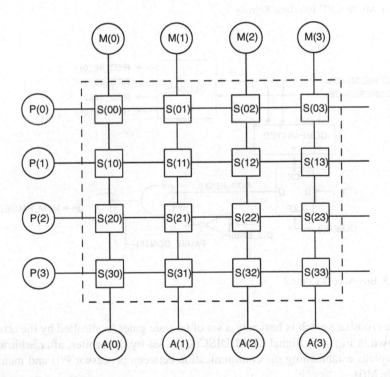

Fig. 4.3 Crossbar Components

wait for the response from the arbiter A(j). A bus access control is, then, implemented, as depicted in Fig. 4.5, which decodifies the two most significant bits of
the current address (ADDRESS(i)<29:28>), generating the corresponding memory module bus request REQ_IN(i,j). If P(i) is requesting its primary bus B(j), for
which $i = j$, the arbiter sets signal PAUSE_CONT(i), pausing the processor until
arbitration is complete, when, then, the arbiter resets this signal. On the other hand,
if P(i) is requesting a secondary bus B(j), for which $i \neq j$, signal BUS_REQ(i) is
set, pausing the processor until arbitration is complete, when, then, the arbiter sets
signal DIS_EN(i), activating the processor.

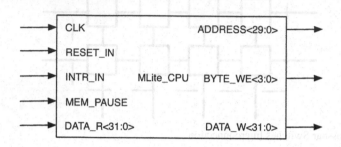

Fig. 4.4 MLite_CPU Interface Signals

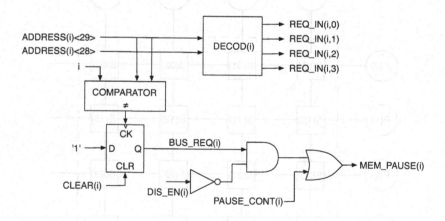

Fig. 4.5 Bus Access Control

The croosbar switch is basically a set of tri-state gates, controlled by the arbiter,
as shown in Fig. 4.6. Signal COM_DISC(i,j) is set by the arbiter, after arbitration
is complete, estabilishing the communication between processor P(i) and memory
module M(j).

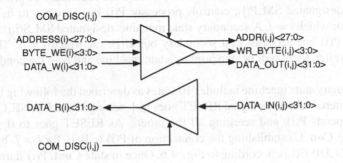

Fig. 4.6 Crossbar Switch

4.2.1 Network Controller

The network controller is composed of the arbiter A(j) and a set of controllers, one for each processor, implemented by state machines SM(j)<0:N-1>, as shown in Fig. 4.7. Upon receiving a bus request, through signals REQ_IN(i,j)<0:N-1>, the arbiter A(j) schedules a processor to be the next bus master, based on the round-robin algorithm, by activating the corresponding signal GRANT(i,j). State machines are used to control the necessary sequence of events to transition from the present bus master to the next one.

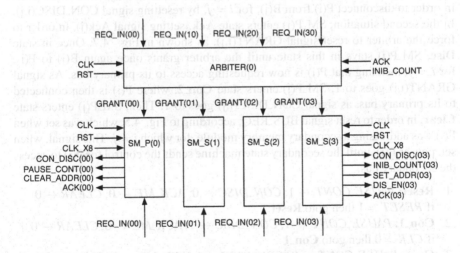

Fig. 4.7 Network Controller

There are two types of state machines: primary and secondary. A primary state machine, designated SM_P(j), controls processor P(i) bus accesses to its primary bus B(j), for which $i = j$. A secondary state machine, designated SM_S(i,j), controls processor P(i) bus accesses to a secondary bus B(j), for which $i \neq j$. Therefore, for each arbiter, there will be one primary state machine and $N - 1$ secondary state machines.

The primary state machine includes 10 states as described the following list. State **Reset** is entered whenever signal RESET goes to 1, setting signal PAUSE_CONT(i), which suspends P(i), and resetting all the others. As RESET goes to 0, SM_P(j) enters state **Con_1**, establishing the connection of P(i) to B(j), for $i = j$, by setting signal CON_DISC(i,j), according to Fig. 4.6. Once in state **Cont**, P(i) starts the bus access, as signal PAUSE_CONT(i) goes to 0. While there are no requests from other processors, so GRANT(i,j)=1 for $i = j$, and P(i) is not requesting any other B(j), so REQ(i,j)=1 for $i = j$, the primary state machine stays in state **Cont**. If another processor requests B(j), for $i = j$, then the arbiter resets signal GRANT(i,j), for $i = j$, and the primary state machine enters state **Pause**, in order to suspend P(i), by setting signal PAUSE_CONT(i). Next, SM_P(j) enters state **P_Disc**, in order to disconnect P(i) from B(j), by resetting signal CON_DISC(i,j). It stays in this state until the arbiter gives B(j) back to P(i), by setting signal GRANT(i,j), for $i = j$. SM_P(j), then, returns to state **Con_1**, where P(i) reestablishes its connection to B(j). On the other hand, from state **Cont**, the other possibility is that P(i) requests another B(j), for $i \neq j$, resetting signal REQ(i,j), for $i = j$. In this case, SM_P(j) enters state **Wait**, in order to check if the arbiter has already granted the secondary bus to P(i), in which case signal GRANT(i,j), for $i = j$, goes to 0, or not yet, in which case signal GRANT(i,j), for $i = j$, remains in 1. In the first situation, SM_P(j) enters state **Disc**, in order to disconnect P(i) from B(j), for $i = j$, by resetting signal CON_DISC(i,j). In the second situation, SM_P(j) enters state **Ack**, setting signal Ack(j), in order to force the arbiter to reset signal GRANT(i,j), as shown in Fig. 4.7. Once in state **Disc**, SM_P(j) stays in this state until the arbiter grants once again B(j) to P(i), for $i = j$, meaning that P(i) is now requesting access to its primary bus. As signal GRANT(i,j) goes to 1, SM_P(j) enters state **Con_2**, where P(i) is then connected to its primary bus, as signal CON_DISC(i,j) goes to 1. Then, SM_P(j) enters state **Clear**, in order to reset signal BUS_REQ, according to Fig. 4.5, which was set when P(i) was addressing a secondary memory module, for which $i \neq j$. This signal, when set, pauses P(i), until the secondary state machine sends the control for P(i) to access the secondary bus.

1. **Reset:** *PAUSE_CONT $\leftarrow$ 1; CON_DISC $\leftarrow$ 0; ACK_ME $\leftarrow$ 0; CLEAR $\leftarrow$ 0*
   if *RESET* $= 1$ then goto **Reset**
2. **Con_1:** *PAUSE_CONT $\leftarrow$ 1; CON_DISC $\leftarrow$ 1; ACK_ME $\leftarrow$ 0; CLEAR $\leftarrow$ 0*
   if *CLK* $= 0$ then goto **Con_1**
3. **Cont:** *PAUSE_CONT $\leftarrow$ 0; CON_DISC $\leftarrow$ 1; ACK_ME $\leftarrow$ 0; CLEAR $\leftarrow$ 0*
   if *GRANT* $= 1$ **and** *REQ* $= 0$ then goto **Wait**
   else if *GRANT* $= 0$ **and** *REQ* $= 1$ then goto **Pause**
4. **Wait:** *PAUSE_CONT $\leftarrow$ 0; CON_DISC $\leftarrow$ 1; ACK_ME $\leftarrow$ 0; CLEAR $\leftarrow$ 0*
   if *GRANT* $= 1$ then goto **ACK_ME** else goto **Disc**

5. **Ack:** $PAUSE\_CONT \leftarrow 0; CON\_DISC \leftarrow 1; ACK\_ME \leftarrow 1; CLEAR \leftarrow 0$
6. **Disc:** $PAUSE\_CONT \leftarrow 0; CON\_DISC \leftarrow 0; ACK\_ME \leftarrow 0; CLEAR \leftarrow 0$
   if $GRANT = 0$ then goto **Disc**
7. **Con_2:** $PAUSE\_CONT \leftarrow 0; CON\_DISC \leftarrow 1; ACK\_ME \leftarrow 0; CLEAR \leftarrow 0$
   if $CLK = 0$ then goto **Con_2**
8. **Clear:** $PAUSE\_CONT \leftarrow 0; CON\_DISC \leftarrow 1; ACK\_ME \leftarrow 1; CLEAR \leftarrow 1$
   goto **Cont**
9. **Pause:** $PAUSE\_CONT \leftarrow 1; CON\_DISC \leftarrow 1; ACK\_ME \leftarrow 0; CLEAR \leftarrow 0$
10. **P_Disc:** $PAUSE\_CONT \leftarrow 1; CON\_DISC \leftarrow 0; ACK\_ME \leftarrow 0; CLEAR \leftarrow 0$
    if $GRANT = 0$ then goto **P_Disc** else goto **CON_1**

The secondary state machine includes 8 states as described by the following list. During initialization, when signal Reset is 1, SM_S(i,j) stays in state **Reset** until the arbiter grants a secondary bus B(j) to processor P(i). When signal GRANT(i,j) goes to 1, SM_S(i,j) enters state **Wait_1**, followed by state **Wait_2**, in order to give time to the corresponding primary state machine to pause P(i) and disconnect it from B(j), for $i = j$. In this case, either P(i) is requesting a secondary bus or another processor is requesting B(j) as secondary bus. In the first situation, SM_P(i,j) enters state **Wait** and in the second situation SM_P(i,j) enters state **Pause**, as discribed above. Observe that only the corresponding signal GRANT(i,j) is set, according to P(i) and B(j) in question. Next, SM_S(i,j) enters state **Con**, where signal CON_DISC(i,j) is set, connecting P(i) to B(j). Once in state **Dis**, signal DIS_EN(i,j) goes to 1, activating P(i), as shown in Fig. 4.5. Recall that P(i), for $i = j$, was paused by the primary state machine, either because it requested a secondary bus or its primary bus is being requested by another processor. Once P(i) finishes using B(j), for which $i \neq j$, SM_S(i,j) enters state **En**, resetting signal DIS_EN(i,j) and pausing P(i). Next, SM_S(i,j) enters state **Disc**, resetting signal CON_DISC(i,j) and disconnecting P(i) from B(j). Then, SM_S(i,j) enters state **Ack**, in order to tell the arbiter it finished using B(j), by setting signal ACK_ME, which makes the arbiter select the next bus master. Observe that signal INIB_COUNT goes to 1 as soon as SM_S(i,j) leaves state **Reset**, stopping the counter that controls the time limit for P(i) to use B(j), for $i = j$, since this processor is not using its primary bus.

1. **Reset:** $CON\_DISC \leftarrow 0; DIS\_EN \leftarrow 0; INIB\_COUNT \leftarrow 0$
   $ACK\_ME \leftarrow 0$; if $GRANT = 0$ then goto **Reset**
2. **Wait_1:** $CON\_DISC \leftarrow 0; DIS\_EN \leftarrow 0; INIB\_COUNT \leftarrow 1; ACK\_ME \leftarrow 0$
3. **Wait_2:** $CON\_DISC \leftarrow 0; DIS\_EN \leftarrow 0; INIB\_COUNT \leftarrow; ACK\_ME \leftarrow 0$
4. **Con:** $CON\_DISC \leftarrow 1; DIS\_EN \leftarrow 0; INIB\_COUNT \leftarrow 1; ACK\_ME \leftarrow 0$
   if $CLK = 0$ then goto **Con**
5. **Dis:** $CON\_DISC \leftarrow 1; DIS\_EN \leftarrow 1; INIB\_COUNT \leftarrow 1; ACK\_ME \leftarrow 0$
   if $CLK = 0$ then goto **Dis**
6. **En:** $CON\_DISC \leftarrow 1; DIS\_EN \leftarrow 0; INIB\_COUNT \leftarrow 1; ACK\_ME \leftarrow 0$
7. **Disc:** $CON\_DISC \leftarrow 0; DIS\_EN \leftarrow 0; INIB\_COUNT \leftarrow 1; ACK\_ME \leftarrow 0$
8. **Ack:** $CON\_DISC \leftarrow 0; DIS\_EN \leftarrow 0; INIB\_COUNT \leftarrow 1; ACK\_ME \leftarrow 1$
   goto **Reset**

4.3 Experimental Results

In order to analyse the performance of the proposed architecture, we used the Particle Swarm Optimization (PSO) method [9][10] to optimize an objective function. This method was chosen due to its intensive computation, being a strong candidate for parallelization. In this method, particles of a swarm are distributed among the processors and, at the end of each iteration, a processor accesses the memory module of another one in order to obtain the best position found in the swarm. The communication between processors is based on three strategies: ring, neighbourhood and broadcast.

4.3.1 Particle Swarm Optimization

The PSO method keeps a swarm of particles, where each one represents a potential solution for a given problem. These particles transit in a search space, where solutions for the problem can be found. Each particle tends to be attracted to the search space, where the best solutions were found. The position of each particle is updated by the velocity factor $v_i(t)$, according to Eq. 4.1:

$$x_i(t+1) = x_i(t) + v_i(t+1) \tag{4.1}$$

Each particle has its own velocity, which drives the optimization process, leading the particle through the search space. This velocity depends on its performance, called cognitive component, and on the exchange of information with its neighbourhood, called social component. The cognitive component quantifies the performance of particle i, in relation to its performance in previous iterations. This component is proportional to the distance between the best position found by the particle, called $Pbest_i$, and its actual position. The social component quantifies the performance of particle i in relation to its neighbourhood. This component is proportional to the distance between the best position found by the swarm, called $Gbest_i$, and its actual position. In Eq. 4.2, we have the definition of the actual velocity in terms of the cognitive and social components of the particle:

$$v_i(t+1) = v_i(t) \times w(t) + c_1 \times r_1(Pbest_i - x_i(t)) + c_2 \times r_2(Gbest_i - x_i(t)) \tag{4.2}$$

Components r_1 and r_2 control the randomness of the algorithm. Components c_1 and c_2 are called the cognitive and social coeficients, controlling the trust of the cognitive and social components of the particle. Most of the applications use $c_1 = c_2$, making both components to coexist in harmony. If $c_1 \gg c_2$, then we have an excessive movement of the particle, making difficul the convergence. If $c_2 \gg c_1$, then we could have a premature convergence, making easy the convergence to a local minimum.

Component w is called the inertia coeficient and defines how the previous velocity of the particle will influence the actual one. The value of this factor is important for the convergence of the PSO. A low value of w promotes a local exploration of the particle. On the other side, a high value promotes a global exploration of the space. In general, we use values near to one, but not too close to 0. Values of w greater than 1 provide a high acceleration to the particle, which can make convergence difficult. Values of w near 0 can make the search slower, yielding an unnecessary computational cost. An alternative is to update the value of w at each iteration, according to Eq. 4.3, where n_{ite} is the total number of iterations. At the beginning of the iterations, we have $w \approx 1$, increasing the exploratory characteristic of the algorithm. During iterations, we linearly decrease w, making the algorithm to implement a more refined search.

$$w(t+1) = w(t) - \frac{w(0)}{n_{ite}} \qquad (4.3)$$

The size of the swarm and the number of iterations are other parameters of the PSO. The first one is the number of existing particles. A high number of particles allows for more parts of the search space to be verified at each iteration, which allows for better solutions to be found, if compared with solutions found in smaller swarms. However, this increases the computational cost, with the increase in execution time. The number of iterations depends on the problem. With few iterations, the algorithm could finish too early, whithou providing an acceptable solution. On the other hand, with a high number of iterations, the computational cost could be unnecessarily high. Algorithm 3.1 describes the PSO method.

4.3.2 Communication between Processes

The parallel execution of the PSO method was done by allocating one instance of the algorithm to each processor of the network. The swarm was then equally divided among the processors. Each subswarm evolves independently and, periodically, *Gbest* is exchanged among the processors. This exchange of data was done based on three strategies: ring, neighbourhood and broadcast.

Fig. 4.8 describes the ring strategy, while Alg. 4.1 describes the PSO using this strategy for process communication. The neighbourhood strategy can be depicted by Fig. 4.9 and the PSO algorithm that implements this strategy is described by Alg. 4.2. Fig. 4.10 shows the broadcast strategy and Alg. 4.3 describes its use for process communication by the PSO algorithm.

4.3.3 Performance Results

The PSO algorithm was used to minimize the Rosenbrock function, defined by Eq. 4.4 and whose curve is shown in Fig. 4.11. We used 1, 2, 4, 8, 16 and 32 processors for each simulation and considering each of the communication strategies, 64 particles, distributed among the processors, and the algorithm was run for 32 iterations. The speedup obtained is described by Fig. 4.12.

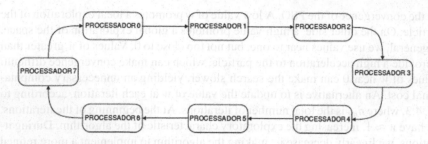

Fig. 4.8 Ring Strategy

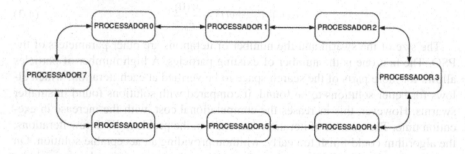

Fig. 4.9 Neighbourhood Strategy

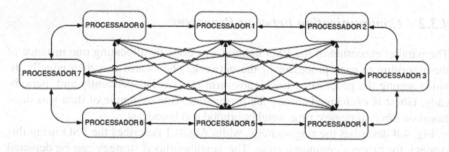

Fig. 4.10 Broadcast Strategy

$$f(x,y) = 100(y - (x^2))^2 + (1 - x)^2 \tag{4.4}$$

4.4 Summary

In order to evaluate the performance offered by the proposed architecture, we executed the PSO method for the minimization of the Rosenbrock function, both sequentially and in parallel. The simulation was done for 1, 2, 4, 8, 16 and 32 processors, using a swarm of 64 particles and implementing 32 iterations. We exploited

Algorithm 4.1. PSO with Ring Strategy

Create and initialize a swarm with n particles
$id := processor identification$
$tmpid := id - 1$
$nproc := number of processors in the network$
if $id \neq 0$ **then**
    $end process(id) := 0$
end if
$tmpid := id - 1$
repeat
    for $j = 1 \rightarrow n$ **do**
        Calculate the fitness of $particle_i$
        Update $Gbest(id)$ and $Pbest(id)$
        Update the particle's velocity
        Update the particle's position
    end for
    Copy $Gbest(id)$ to share the memory
    Read $Gbest$ from $processor(tmpid)$
    if $Gbest(tmpid) \leq Gbest(id)$ **then**
        $Gbest(id) := Gbest(tmpid)$
    end if
until Stop criteria = true
if $id = 0$ **then**
    $Best := Gbest(id)$
    $tmpid := id + 1$
    for $k = 1 \rightarrow nproc - 1$ **do**
        Read $end process(tmpid)$
        while $end process(tmpid) = 0$ **do**
            Read $end process(tmpid)$
        end while
        Read $Gbest$ from $processor(tmpid)$
        if $Gbest(tmpid) \leq Best$ **then**
            $Best := Gbest(tmpid)$
        end if
        $tmpid := tmpid - 1$
    end for
else
    $end process(id) := 1$
end if

---

three communication strategies: ring, neighbourhood and broadcast. The speedup obtained demonstrated that the performance offered by the network increases with the number of processors. Another fact is that both ring and neighbourhood strategies have similar impact on the performance of the network, while the broadcast strategy decreases the performance. This decrease is due to the fact the the latter imposes much more interprocess communication than the former ones.

Algorithm 4.2. PSO with Neighborhood Strategy

Create and initialize a swarm with n particles
$id := processor identification$
$tmpid := id - 1$
$nproc := number of processors in the network$
if $id \neq 0$ **then**
    $endprocess(id) := 0$
end if
repeat
    for $j = 1 \rightarrow n$ **do**
        Calculate the fitness of $particle_i$
        Update $Gbest(id)$ and $Pbest(id)$
        Update the particle's velocity
        Update the particle's position
    end for
    Copy $Gbest(id)$ to share the memory
    $tmpid := id + 1$
    Read $Gbest$ from $processor(tmpid)$
    if $Gbest(tmpid) \leq Gbest(id)$ **then**
        $Gbest(id) := Gbest(tmpid)$
    end if
    $tmpid := id - 1$
    Read $Gbest$ from $processor(tmpid)$
    if $Gbest(tmpid) \leq Gbest(id)$ **then**
        $Gbest(id) := Gbest(tmpid)$
    end if
until Stop criteria = true
if $id = 0$ **then**
    $Best := Gbest(id)$
    $tmpid := id + 1$
    for $k = 1 \rightarrow nproc - 1$ **do**
        Read $endprocess(tmpid)$
        while $endprocess(tmpid) = 0$ **do**
            Read $endprocess(tmpid)$
        end while
        Read $Gbest$ from $processor(tmpid)$
        if $Gbest(tmpid) \leq Best$ **then**
            $Best := Gbest(tmpid)$
        end if
        $tmpid := tmpid + 1$
    end for
else
    $endprocess(id) := 1$
end if

Algorithm 4.3. PSO with Broadcast Strategy

Create and initialize a swarm with n particles
$id := processoridentification$
$nproc := numberofprocessorsinthenetwork$
if $id \neq 0$ **then**
    $endprocess(id) := 0$
end if
repeat
    for $j = 1 \rightarrow n$ **do**
        Calculate the fitness of $particle_i$
        Update $Gbest(id)$ and $Pbest(id)$
        Update the particle's velocity
        Update the particle's position
    end for
    Copy $Gbest(id)$ to share the memory
    $tmpid := id + 1$
    for $k = 1 \rightarrow nproc - 1$ **do**
        Read $Gbest$ from $processor(tmpid)$
        if $tmpid = nproc - 1$ **then**
            $tmpid = 0$
        else
            $tmpid := tmpid + 1$
        end if
    end for
until Stop criteria = true
if $id = 0$ **then**
    $Best := Gbest(id)$
    $tmpid := id + 1$
    for $k = 1 \rightarrow nproc - 1$ **do**
        Read $endprocess(tmpid)$
        while $endprocess(tmpid) = 0$ **do**
            Read $endprocess(tmpid)$
        end while
        Read $Gbest$ from $processor(tmpid)$
        if $Gbest(tmpid) \leq Best$ **then**
            $Best := Gbest(tmpid)$
        end if
        $tmpid := tmpid + 1$
    end for
else
    $endprocess(id) := 1$
end if

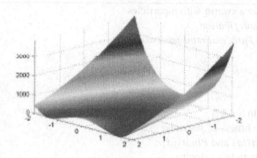

Fig. 4.11 Graphic of the Rosenbrock Function

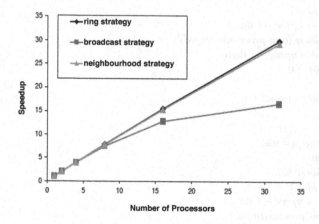

Fig. 4.12 Speedup obtained for the Execution of the Rosenbrock Function

As for future work, we intend to explore other applications for parallelization, in order to analyse the impact of their behaviour specially concerning the interprocess communication; introduce cache memory, to improve performance; develop a microkernel, to implement task scheduling and explore multithread execution; explore other arbitration schemes; sintezise the architecture, in order to analyse the cost x performace relation.

References

1. Kongerita, P., et al.: Niagara: 32-way multithreaded Spark processor. IEEE MICRO 25(2), 21–29 (2005)
2. Freitas, H.C.: NoC Architecture Design for Multi-Cluster Chips. In: IEEE International Conference on Field Programmable Logic and Applications, pp. 53–58. IEEE Press, New York (2008)

3. Patterson, D.A., Hennessy, J.L.: Computer Organization: The Hardware/Software Interface, 3rd edn. Morgan Kaufmann, San Francisco (2005)
4. Pande, P.T., Michele, G., et al.: Design, Synthesis, and Test of Networks on Chips. IEEE Design & Test of Computers (2005)
5. Matt, W.: Arbiters: Design Ideas and Coding Styles. Silicon Logic Engineering, Inc. (2001)
6. Duato, J., Yalamanchili, S., Ni, L.: Interconnection Networks: An Engineering Approach. Morgan Kaufmann, San Francisco (2003)
7. Ni, L.M.: Issues in Designing Truly Scalable Interconnection Networks. In: International Conference on Parallel Processing Workshop, pp. 74–83. IEEE Press, New York (1996)
8. OpenCores, http://www.opencores.org
9. Kennedy, J., Eberhart, R.: Particle swarm optimization. In: IEEE International Conference on Neural Networks, vol. 4, pp. 1942–1948. IEEE Press, New York (1995)
10. Engelbrecht, A.P.: Fundamentals of Computational Swarm Intelligence. John Wiley & Sons, Chichester (2006)
11. Tanenbaum, A.S.: Structured Computer Organization, 5th edn. PEARSON Prentice Hall, New Jersey (2006)

Chapter 5
A Reconfigurable Hardware for Artificial Neural Networks*

Abstract. Artificial Neural Networks (ANNs) is a well known bio-inspired model that simulates human brain capabilities such as learning and generalization. ANNs consist of a number of interconnected processing units, wherein each unit performs a weighted sum followed by the evaluation of a given activation function. The involved computation has a tremendous impact on the implementation efficiency. Existing hardware implementations of ANNs attempt to speed up the computational process. However these implementations require a huge silicon area that makes it almost impossible to fit within the resources available on a state-of-the-art FPGAs. In this chapter, we devise a hardware architecture for ANNs that takes advantage of the dedicated adder blocks, commonly called MACs to compute both the weighted sum and the activation function. The proposed architecture requires a reduced silicon area considering the fact that the MACs come for free as these are FPGA's built-in cores. The hardware is as fast as existing ones as it is massively parallel. Besides, the proposed hardware can adjust itself on-the-fly to the user-defined topology of the neural network, with no extra configuration, which is a very nice characteristic in robot-like systems considering the possibility of the same hardware may be exploited in different tasks.

5.1 Introduction

Artificial Neural Networks (ANNs) are useful for learning, generalization, classification and forecasting problems [3]. They consists of a pool of relatively simple processing units, usually called artificial neurons, which communicates with one another through a large set of weighted connections. There are two main network topologies, which are feed-forward topology [3], [4] where the data flows from input to output units is strictly forward and recurrent topology, where feedback connections are allowed. Artificial neural networks offer an attractive model that allows one to solve hard problems from examples or patterns. However, the computational process behind this model is complex. It consists of massively parallel non-linear

---

* This chapter was developed in collaboration with Rodrigo Martins da Silva.

N. Nedjah and L. de Macedo Mourelle, *Hardware for Soft Computing and Soft Computing for Hardware*, Studies in Computational Intelligence 529,
DOI: 10.1007/978-3-319-03110-1_5, © Springer International Publishing Switzerland 2014

calculations. Software implementations of artificial neural networks are useful but hardware implementations takes advantage of the inherent parallelism of ANNs and so should answer faster.

Field Programmable Gate Arrays (FPGAs) [7] provide a re-programmable hardware that allows one to implement ANNs very rapidly and at very low-cost. However, FPGAs lack the necessary circuit density as each artificial neuron of the network needs to perform a large number of multiplications and additions, which consume a lot of silicon area if implemented using standard digital techniques.

The proposed hardware architecture described throughout this chapter is designed to process any fully connected feed-forward multi-layer perceptron neural network (MLP). It is now a common knowledge that the computation performed by the net is complex and consequently has a huge impact on the implementation efficiency and practicality. Existing hardware implementations of ANNs have attempted to speed up the computational process. However these designs require a considerable silicon area that makes tem almost impossible to fit within the resources available on a state-of-the-art FPGAs [1], [2], [6]. In this chapter, we devise an original hardware architecture for ANNs that takes advantage of the dedicated adder blocks, commonly called MACs (short for Multiply, Add and Accumulate blocks) to compute both the weighted sum and the activation function. The latter is approximated by a quadratic polynomial using the least-square method. The proposed architecture requires a reduced silicon area considering the fact that the MACs come for free as these are FPGA's built-in cores. The hardware is as fast as existing ones as it is massively parallel. Besides, the proposed hardware can adjust itself on-the-fly to the user-defined topology of the neural network, with no extra configuration, which is a very nice characteristic in robot-like systems considering the possibility of the same piece of hardware may be exploited in different tasks.

The remaining of this chapter is organized as follows: In Section 5.2, we give a brief introduction to the computational model behind artificial neural networks; In Section 5.3, we show how we approximate the sigmoid output function so we can implement the inherent computation using digital hardware; In Section 5.4, we provide some hardware implementation issues about the proposed design, that makes it original, efficient and compact; In Section 5.5, we present the detailed design of the proposed ANN Hardware; Last but no least, In Section 5.6, we draw some useful conclusions and announce some orientations for future work.

5.2 ANNs Computational Model

We now give a brief introduction to the computational model used in neural networks. Generally, is constituted of few layers, each of which includes several neurons. The number of neurons in distinct layers may be different and consequently the number of inputs and that of outputs may be different [3].

The model of an artificial neuron requires n inputs, say $I_1, I_2, \ldots, I_n$ and the synaptic weights associated with these inputs, say $w_1, w_2, \ldots, w_n$. The weighted sum a, which, also called activation of the neuron, is defined in (5.1). The model usually includes an output function $nout(.)$ that is applied to the neuron activation before it is fed forwardly as input to the next layer neurons.

$$a = \sum_{j=1}^{n} w_j \times f_j \qquad (5.1)$$

The non-linearity of the neuron is often achieved by the output function, which may be the hyperbolic tangent or sigmoid [3]. In some cases, $nout(a)$ may be linear.

A typical ANN operates in two necessary stages: *learning* and *feed-forward computing*. The learning stage consists of supplying known patterns to the neural network so that the network can adjust the involved weights. Once the network has learned to recognize the provided patterns, the network is ready to operate, performing the feed-forward computing. In this stage, the network is supplied with an input data or pattern, which may or not be one of those given in learning stage and verify how the network responds with output results. This allows one to know whether the neural network could recognize the input data. The precision of the net in recognizing the new input patterns depends on the quality of its learning stage and on its generalization. As we have previously mentioned, here we are only concerned with the implementation of feed-forward computing stage.

5.3 Approximation of the Output Function

Unlike the activation function, which includes operations that can easily and efficiently implemented in hardware, the out function requires a special care before the computation involved can be modeled in hardware. Without loss of generality, we chose to use the sigmoid output function. Note that the same treatment applies to the hyperbolic function too. To allow an efficient implementation of the sigmoid function defined in (5.2), in hardware, we proceeded with a parabolic approximation of this function using the least-square estimation method.

$$sigmoid(a) = \frac{1}{1+e^{-a}} \qquad (5.2)$$

The approximation proceeds by defining $nout(a) = C \times a^2 + B \times a + A$ as a parabolic approximation of the sigmoid of (5.2), just for a short range of the variable a. We used the least-square parabola to make this approximation feasible. Many attempts were performed to try to find out the best range of a for this approximation, so that the parabola curves fits best that of $sigmoid(a)$. We obtained the range $[-3.3586, 2.0106]$ for variable a, taking into account the calculated coefficients $C = 0.0217$, $B = 0.2155$ and $A = 0.4790$ for the parabolic approximation. Thus, the approximation of the sigmoid function is as defined in (5.3):

$$nout(a) = \begin{cases} 0 & a < -3.3586 \\ 0.0217 \times a^2 + 0.2155 \times a + 0.4790 & a \in [-3.3586, 2.0106] \\ 1 & a > 2.0106 \end{cases} \qquad (5.3)$$

5.4 Implementation Issues

An Artificial Neural Network is a set of several interconnected neurons arranged in layers. Let L be the number of layers. Each layer has its own number of neurons. Let m_i be the number of neurons in layer i. The neurons are connected by the synaptic connections. Some neurons get the input data of the network, so they are called *input* neurons and thus compose the input layer. Other neurons export their outputs to the outside world, so these are called output neurons and thus compose the output layer. Neurons placed on the layer 2 up to layer $L-1$ are called the hidden neurons because they belong to the hidden layers. In Fig. 5.1, we show a simple example of an ANN. The output of each neuron, save output neurons, represents an input of all neurons placed in the next layer.

The computation corresponding to a given layer starts only when that of the corresponding previous layer has finished. Our ANN hardware has just one *real* layer of neurons, constitutes of k neurons, where k is maximum number of neurons per layer, considering all layers of the net. For instance, for the net of Fig. 5.1, this parameter is 3. This single real layer or physical layer is used to implement all layers of the network. As only one layer operates at a time, this allows us to minimize drastically the silicon area required without altering the response time of the net. For instance, considering the net of Fig. 5.2, the first stage of the computation would use only 2 neurons, then in the second stage all three physical neurons would be exploited and in the last stage, only one neuron would be useful. So instead of having 6 physically implemented neurons, our hardware requires only half that number to operate. ANN hardware treats the nets layers as virtual.

Besides reducing the number of neurons that are actually implemented in hardware, our design takes advantage of some built-in cores that come for free in nowadays FPGAs. This blocks are called MACs (Multiply, add and Accumulate), which are usually used in DSPs (Digital Signal Processing) and their architecture is shown in Fig. 5.2. The MACs blocks are perfect to perform the weighted sum.

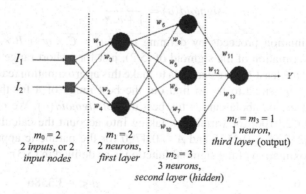

Fig. 5.1 Neural network with two inputs, three layers and one output Y

Recall that $nout(a)$ of (5.3) is the actual neuron output function our ANN hardware will perform. Observe that the computation involved in this function is sum of products (quadratic polynomial) and so the MACs can be used in this purpose to. Actually we use the same block of the neuron to compute the output function.

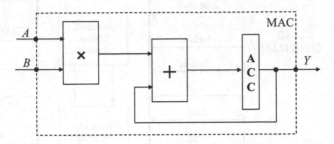

Fig. 5.2 Built-in MACs blocks in FPGAs

5.5 ANN Hardware Architecture

The ANN hardware interface is illustrated in Fig. 5.3, wherein two other components are included: LOAD CONTROLLER and CLOCK SYSTEM. The former may be any outside system able to setup the neural network topology and to load all necessary data in the ANN hardware. This include the number of inputs, the number of layers, the number of neurons per layer and that of outputs, besides, the net inputs and the definitive weights. Our ANN hardware is organized in a neural control unit (UC) and Neural arithmetic and logic unit (ALU).

Neural UC encompasses all control components for computing all neural network feed-forward computation. It also contains the memories for storing the net's inputs in the INPUT MEMORY, the weights in the WEIGHT MEMORY, the number of inputs and neurons per layer in the LAYER MEMORY and the coefficients of the output function in the OUTPUT FUNCTION MEMORY as described in (5.3). Fig. 5.4 and Fig. 5.5 depict, respectively, two parts of the neural UC.

During the loading process, which commences when $LCStart = 1$, the LOAD CONTROLLER sets signal *DataLoad* and selects the desired memory of the neural UC by signals $Load_0$ and $Load_1$ (see Fig. 5.3, Fig. 5.4 and Fig. 5.5).

The counters that provide addresses for memories are entirely controlled by the LOAD CONTROLLER. Signal *JKClk* is the clock signal (from CLOCK SYSTEM in Fig. 5.4) that synchronizes the actions of those Counters and of the LOAD CONTROLLER. This one fills each memory through the 32-bit *DATA* in loading process.

When the loading process is finished ($LCFinal = 1$), in Fig. 5.3, signal *DataLoad* can be turned off and the LOAD CONTROLLER can set signal *Start* for the commencing of the feed-forward Neural Network computing. When $Start = 1$ (and $DataLoad = 0$), the ANN hardware gets the whole control of its components; so the LOAD CONTROLLER can no longer interfere in the neural UC. This one has a

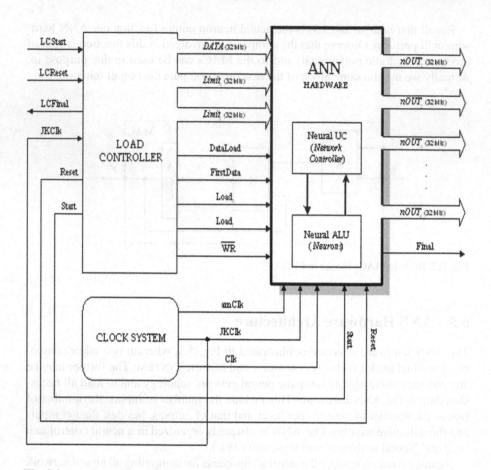

Fig. 5.3 Interface of the ANN hardware

main controller called *Network Controller* (Fig. 5.3) that controls all components of the neural UC (Fig. 5.4 and Fig. 5.5) and also the neural ALU, which is depicted in Fig. 5.6.

During the ANN hardware operation, neural UC by the mean of the network controller, controls the computation of each layer per stage. For each layer of the neural network, all k hardware neurons of the neural ALU of Fig. 5.6 work in parallel even though not necessarily all physical neurons are needed in the layer. Recall that some layers in the ANN hardware may have fewer neurons than k. At this stage, signal *Clk* is now the active clock of the ANN hardware, not signal *JKClk* anymore.

In Fig. 5.6, ADDER MUX decides the actual input for all hardware neurons and it is exploited to multiplex a network input, from the INPUT MEMORY in Fig. 5.4 or the output of a hardware neuron $nOUT_i$, which is an output of a neuron placed in a layer i of the net. While all physical neurons are in operation, the WEIGHT REGISTERS of Fig. 5.6 are already being loaded using signal W (see Fig. 5.4). These are the new

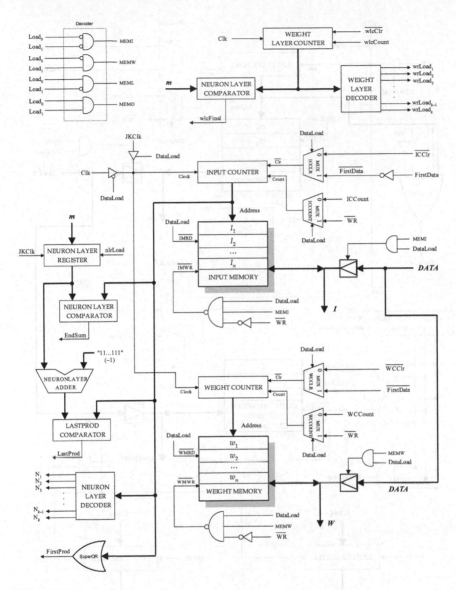

Fig. 5.4 First part of the neural UC

weights, which are the weights of the next layer in the network. Furthermore, in Fig.
5.6, we see a set of tri-state buffers, each of which is controlled by signal N_i, issued
by the NEURON LAYER DECODER, in the neural UC of Fig. 5.4. Fig. 5.6 shows the
neuron architecture. Each hardware neuron performs the weighted sum followed by
the application of the output function $nout(a)$.

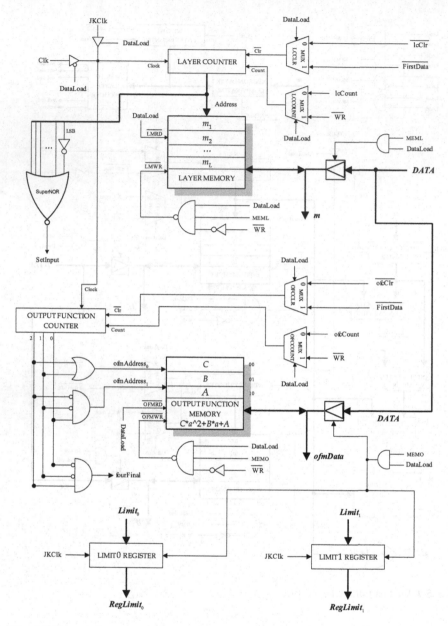

Fig. 5.5 Second part of the neural UC

Observing Fig. 5.4 (neural UC), the INPUT COUNTER, together with NEURON LAYER REGISTER, NEURON LAYER COMPARATOR, NEURON LAYER ADDER and LASTPROD COMPARATOR control the computation of the weighted sum: signal *FirstProd* indicates the first product of the weighted sum and *LastProd*, the last

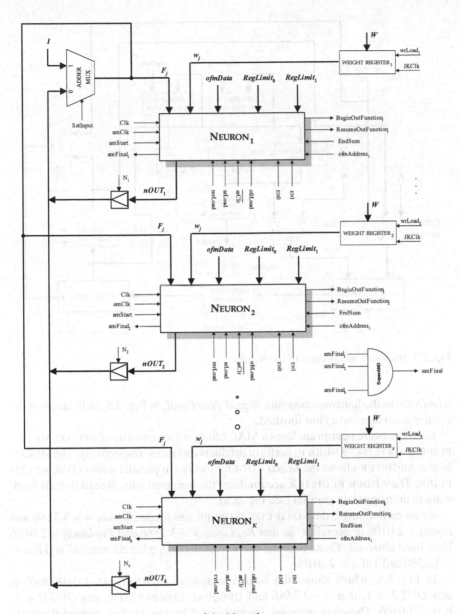

Fig. 5.6 Overall hardware architecture of the Neural ALU

one. SuperOR component is an OR of all input bits. Signal *EndSum* (Fig. 5.4, Fig. 5.5 and Fig. 5.6) flags that the weighted sum has been completed. It also triggers the start of the output function computation. During this stage, the OUTPUT FUNCTION COUNTER (see Fig. 5.5) provides the address to the OUTPUT FUNCTION MEMORY in order to release the coefficients ($C = 0.0217$, $B = 0.2155$ or $A = 0.4790$), through

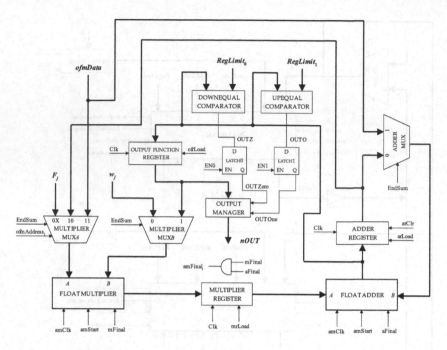

Fig. 5.7 Hardware architecture of the Neuron

$ofmData$, to the hardware neurons. Signal $fourFinal$, in Fig. 5.5, indicates that the computation of $nout(a)$ has finished.

Each hardware neuron encloses a MAC block, which consists of a FLOAT MULTI-PLIER and a FLOAT ADDER to perform products and sums, respectively. The MULTI-PLIER REGISTER allows the LOAD ADDER to works in parallel with FLOAT MULTI-PLIER. The ADDER REGISTER accumulates the weighted sum. Recall that all hardware neurons work in parallel (see Fig. 5.6).

At an earlier stage, the LOAD CONTROLLER has loaded $Limit_0 = -3.3586$ and $Limit_1 = 2.0106$ in neural UC so that $RegLimit_0 = -3.3586$ and $RegLimit_1 = 2.0106$ have been obtained. Those float numbers refer to (5.3), wherein $nout(a)$ is 0 if $a < -3.3586$ and 1 if $a > 2.0106$.

In Fig. 5.6, which shows the hardware neuron, DOWNEQUAL COMPARATOR sets $OUTZ = 1$, if $a < -3.3586$ and UPEQUAL COMPARATOR sets $OUTO = 1$, if $a > 2.0106$. These components, intermediated by two latches, control the OUT-PUT MANAGER, which decides as to the output of the hardware neuron ($nOUT$): *(i)* If $a \in [-3.3586, 2.0106]$, then $nOUT$ is the result of the second degree polynomial as described in (5.3), which is the content of the OUTPUT FUNCTION REG-ISTER; *(ii)* If $a < -3.3586$, then the OUTPUT MANAGER provides $nOUT = 0$; *(iii)* If $a > 2.0106$, then $nOUT$ is 0. Components LATCH$_0$ and LATCH$_1$ are used to maintain $nOUT$ stable. Signal $nOUT$ have to be kept during the computation of the weighted sum of a next layer neuron. Furthermore, in Fig. 5.6, signal $amFinal_i$

indicates the end of both a product and sum performed by the neuron. The multiplier and the adder operate in parallel, i.e. when the adder is accumulating the freshly computed product to the partial weighted sum obtained so far, the multiplier is computing the next product. In Fig. 5.5, signal *amFinal* indicates the end of all the computation in all neurons.

In Fig. 5.3, signal *Final* indicates that all computation required in all the layers of the network are completed and the outputs of the network have been obtained. These outputs are available signals $nOUT_1$, $nOUT_2$, ..., $nOUT_h$ (see Fig. 5.3 and 5.7), where h is the number of neurons placed in the output layer of the Network, with $h \leq k$.

5.6 Summary

In this chapter, we presented novel hardware architecture for processing an artificial neural network, whose topology can be changed on-the-fly without any extra reconfiguration effort. The design takes advantage of the built-in MACs block that come for free in modern FPGAs. The model was specified in VHDL [5], simulated to validate its functionality. We are now working on the synthesis process to evaluate time and area requirements. The comparison of the performance result of our design will be then compared to both the binary-radix straight forward design and the stochastic computing based design.

References

1. Bade, S.L., Hutchings, B.L.: FPGA-Based Stochastic Neural Networks Implementation. In: IEEE Workshop on FPGAs for Custom Computing Machines, pp. 189–198. IEEE Press, Napa (1994)
2. Brown, B.D., Card, H.C.: Stochastic Neural Computation II: Soft Competitive Learning. IEEE Transactions on Computers 50(9), 906–920 (2001)
3. Hassoun, M.H.: Fundamentals of Artificial Neural Networks. MIT Press, Cambridge (1995)
4. Moerland, P., Fiesler, E.: Neural Network Adaptation to Hardware Implementations. In: Fiesler, E., Beale, R. (eds.) Handbook of Neural Computation, Oxford, New York (1996)
5. Navabi, Z.: VHDL: Analysis and Modeling of Digital Systems, 2nd edn. McGraw Hill (1998)
6. Nedjah, N., Mourelle, L.M.: Reconfigurable Hardware for Neural Networks: Binary radix vs. Stochastic. Journal of Neural Computing and Applications 16(3), 249–255 (2007)
7. Xilinx, Inc. Foundation Series Software, http://www.xilinx.com

Chapter 6
A Reconfigurable Hardware for Fuzzy Controllers*

Abstract. Computational system modeling is full of ambiguous situations, wherein the designer cannot decide, with precision, what should be the outcome of the system. Process control is one of the many applications that took advantage of the fuzzy logic. Controller are usually embedded into the controller device. This chapter aims at presenting the development of a reconfigurable efficient architecture for fuzzy controllers, suitable for embedding. The architecture is parameterizable so it allows the setup and configuration of the controller so it can be used for various problem applications. An application of fuzzy controllers was implemented and its cost and performance are presented.

6.1 Introduction

In [18], L. Zadeh introduced for the first time the concept of *fuzziness* as opposed to *crispiness* in data sets. When he invented *fuzzy sets* together with the underlying theory, Zadeh's main concern was to reduce system complexity and provide designer with a new computing paradigm that allow the to approximate results. Whenever there is uncertainty, *fuzzy logic* together with *approximate reasoning* apply. Fuzzy logic and approximate reasoning [17] can be used in system modeling and control as well as data clustering and prediction, to name only few appropriate applications. Furthermore, they can be applied to any discipline such as finance, image processing, temperature and pressure control, robot control, among many others. The Fuzzy Logic is a subject of great interest in scientific circles, but it is still not commonly used in industry, as it should be. Eventually, we found some literature containing practical applications that is being currently used in industry [11, 13].

Fuzzy logic has been used in many of applications, such as expert systems, computing with words, approximate reasoning, natural language, process control, robotics, modeling partially open systems, pattern recognition, decision making and data clustering [12].

---

* This chapter was developed in collaboration with Paulo Renato de Souza e Silva Sandres.

N. Nedjah and L. de Macedo Mourelle, *Hardware for Soft Computing and Soft Computing for Hardware*, Studies in Computational Intelligence 529,
DOI: 10.1007/978-3-319-03110-1_6, © Springer International Publishing Switzerland 2014

There are many related works that implemented a fuzzy controller on a FPGA, but most of them present controller designs that are only suitable for a specific application [9] [13]. Mainly, the designs do not use 32-bit floating-point data. The floating-point data representation is crucial for the sensibility of the controller design. In contrast, all the required computation in the proposed controller are performed by a simple precision floating-point co-processor.

The purpose of the development of a reconfigurable hardware of a *shell* fuzzy controller, that can include any number of inputs and outputs as well as any number of rules, is the possibility of creating a device that can be used more widely and perhaps spread the concept of fuzzy logic in the industrial final products.

This paper is divided into three sections. First, in Section 6.2, we introduce briefly some concepts of fuzzy controller, which will be useful to follow the description of the proposed architecture. Then, in Section 6.3, we describe thoroughly, the macro-architecture of the fuzzy controller developed. After that, in Section 8.3, we give details about the main components included in the macro-architecture. Subsequently, in Section 9.8, we show, via the project of two fuzzy controller, that the proposed architecture is functionally operational and promising in terms of cost and performance. Finally, in Section 8.5, we draw some conclusions and point out some new direction for the work in progress.

6.2 Fuzzy Controllers

Fuzzy control, which directly uses fuzzy rules, is the most important and common application of the fuzzy theory [16]. Using a procedure originated by E. Mamdani [11], three steps are followed to design a fuzzy controlled machine:

1. fuzzification or encoding: This step in the fuzzy controller is responsible of encoding the crisp measured values of the system parameter into a fuzzy term using the respective membership functions;
2. inference: This step consists of identifying the subset of fuzzy rules that can be fired, i.e. those with antecedent propositions with truth degree not zero, and draw the adequate fuzzy conclusions;
3. defuzzification or decoding: This is the reverse process of fuzzification. It is responsible of decoding a fuzzy variable and compute its crisp value.

The generic architecture of a fuzzy controller is given in Fig. 6.1. The main components of a fuzzy controller consist of a *knowledge repository*, the *encoder* or *fuzzification* unit, the *decoder* or *defuzzification* unit and the *inference engine*. The knowledge base stores two kind of data: the fuzzy rules which are required by the inference engine to reach the expected results and knowledge about the fuzzy terms together with their respective membership functions as well as information about the universe of discourse of each fuzzy variable manipulated within the controller. The encoder implements the transformation from crisp to fuzzy and the decoder

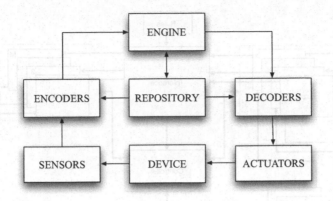

Fig. 6.1 Generic architecture of fuzzy controllers

the transformation from fuzzy to crisp. Of course, the inference engine is the main component of the controller architecture. It implements the approximate reasoning process.

6.3 The Proposed Macro-architecture

The macro-architecture of the proposed fuzzy controller consists of three main units: *(i)* the fuzzification unit (FU), which is responsible for translating the input values of the system into fuzzy terms using the respective membership functions. This unit has as many Fuzzy blocks as required in fuzzy system model that is being implemented, i.e. one for each input variable; *(ii)* the inference unit Inference, which checks all the included fuzzy rules, verifying which membership function applies, and if any is so, generating its value and thus identifying the membership functions to be used in the sequel; *(iii)* the defuzzification unit (DU), which is responsible for translating the fuzzy terms back so as to compute the crisp value of the fuzzy controller output. The defuzzification unit includes as many Defuzzy blocks as required by the fuzzy system model that is being implemented, i.e. one for each output variable. The block diagram of the proposed macro-architecture is shown in the Fig. 6.2, wherein N and M represent the number of input and output variables, respectively.

Note that, besides the main units, the macro-architecture also includes a component that allows to compute the membership function characteristics, which are used by both the fuzzification and defuzzification units. This component will be called membership function unit (MFU). It includes as many MF blocks as required input variable of the fuzzy model. Note that all the membership function-related data are stored in the membership function memory, called MF MEM. This memory is formed by as many memory segments as required input variables, i.e. one for each membership function used. The rules used by the inference unit are stored in a read-only memory block, called Rules. Component Controller, which in the

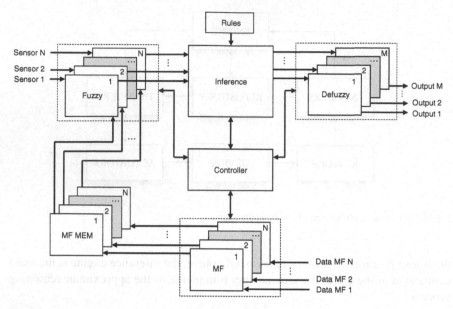

Fig. 6.2 Macro-architecture of the designed fuzzy controller

sequel may be called *main controller*, imposes the necessary sequencing and/or the simultaneity of the required steps of the fuzzy controller via a concurrent finite state machine. More details on this are given subsequently.

The proposed fuzzy controller is designed to be generic and parametric, so it allows configuring the number of input and output variables, the number of linguistic terms used to model the membership functions and the number of inference rules, so as the fuzzy system model that is being implemented can fit in. Allowing the configuration of these parameters makes it possible, as well as easy, to adjust the controller design to any desired problem. Summing up, the main parameters of the controller are:

- N: The number of input variables and hence that of the included Fuzzy blocks;
- M: The number of output variables and hence that of Defuzzy blocks;
- P: The number of rules and thus the number of words in the rule base Rules;
- Q: The number of linguistic terms per membership functions used to model the input and output variables of the fuzzy system.

As it can be seen in the Fig. 6.2, at configuration time, all the membership functions used by the controller are computed and stored in the respective MF MEM segment of the membership function memory. All the computed data will be readily available to be used by the pertinent Fuzzy and/or Defuzzy block in the fuzzification and defuzzification unit, respectively. Note that this configuration step is done only once. During the operation step, the fuzzy controller iterates the required steps, triggering the Fuzzy blocks then Inference unit then Defuzzy blocks in sequence. After that, it waits for a new set of input data to be read by the system sensors and thus arrive at the `Fuzzy` blocks input ports. The finite state machines that control the `Fuzzy` blocks all run in parallel, so do those that control the `Defuzzy` blocks.

In the following sections of this chapter, more light will be shed on the internal micro-architecture of the proposed design as well as the control used therein.

6.4 Micro-architecture of the Functional Units

In this section, we describe the micro-architecture of the main components, included in the macro-architecture of Fig. 6.2. These are the functional unit responsible for the computation of the member function (`MF`), including the memory-based component (`MF MEM`), the basic component responsible for the fuzzification process (`Fuzzy`), the component that implements the inference process (`Inference`) using the available rule base (`Rules`) and the basic component that handles the defuzzification process (`Defuzzy`). In general, all blocks that perform floating-point computations include an `FPU` unit, which performs the main mathematical operations with simple precision (32 bits). The operations needed are addition, subtraction, multiplication and division.

6.4.1 Membership Function Unit

A membership function is viewed as a set of linguistic terms, each of which is defined by two straight lines. In the proposed design, the triangular shape is used to represent linguistic terms. Nevertheless, it is possible to adjust the design as to accept other used shapes such as trapezes and sigmoid. Fig. 6.3 shows a generic example of membership function with Q linguistic terms, wherein the horizontal axis x represents the controller's input, probably read from a sensor, and the vertical axis y represents the truth degree associated with the linguistic terms. This is a real value, between 0 and 1, handled as a simple precision floating-point number of 32 bits. Linguistic terms of triangular membership function are completely defined by *MaxPoint* or *mp* and *Range* or *r*, as illustrated Fig. 6.3.

The `MF` block is designed to compute the values of any variable x, according to $y = ax + b$ of the two straight lines, that represent the linguistic term of the membership function. The required basic data that completely define these shapes need to be identified.

The input data of the `MF` block are *MaxPoint* and *Range* for each straight line used to define the linguistic terms of the membership function. The block utilizes them and pre-compute coefficients a and b accordingly and stored them in the membership function memory segments. Three cases are possible: the leftmost linguistic term (see linguistic term 0 in Fig. 6.3); An in-between linguistic term (see linguistic term 1 and 2 in Fig. 6.3); and finally, the rightmost linguistic term (see linguistic term Q in Fig. 6.3). The computation of a and b of the straight lines of the

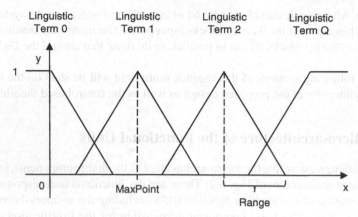

Fig. 6.3 Membership function of Q linguistic terms

leftmost, middle and rightmost linguistic terms are defined as in (6.1), (6.2) and (6.3), respectively.

$$\mu_l(x) = \begin{cases} 1, & \text{if } x \leq mp \\ -\frac{1}{r}x + \frac{mp}{r} + 1, & \text{if } mp > x \geq mp + r \\ 0, & \text{otherwise} \end{cases} \quad (6.1)$$

$$\mu_m(x) = \begin{cases} \frac{1}{Range} \times x - \frac{mp-r}{r}, & \text{if } mp - r < x \leq mp \\ -\frac{x}{r} + \frac{mp}{r} + 1, & \text{if } mp > x \geq mp + r \\ 0, & \text{otherwise} \end{cases} \quad (6.2)$$

$$\mu_r(x) = \begin{cases} \frac{1}{r} \times x - \frac{mp-r}{r}, & \text{if } mp - r < x \leq mp \\ 1, & \text{if } x > mp \\ 0, & \text{otherwise} \end{cases} \quad (6.3)$$

The micro-architecture of the membership function blocks MF is shown in Fig. 6.4. It uses a floating-point unit to perform the required mathematical operations. The obtained results are then stored in the MF MEM segments.

An MF block includes a controller that is implemented as a finite state machine. It allows to synchronize the setting up of all the linguistic terms, necessary to the complete definition of the membership function for each input variable.

In the following, we sketch how the membership function block works. When and MF block receives the enable command from the main controller, the state machine of Fig. 6.5 transits from state *start* to *test_step*, where it checks whether this stage

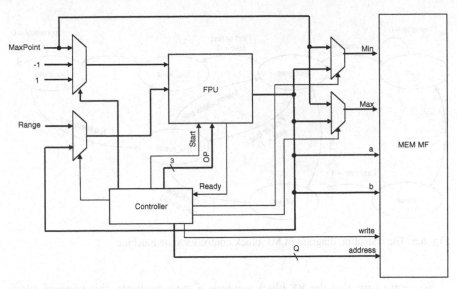

Fig. 6.4 The micro-architecture of the membership function block

is the first or the last straight line calculation of the membership function. If this is the case, there is no need to do anything else, because the first and last straight lines are constants, as it can be seen in Fig. 6.3. Therefore, the result of FPU block is ignored and the FSM goes to state *fpu_result*. Otherwise, i.e. if this is not the first or last straight line, the FSM transits to state *fpu_load*, wherein the values of *MaxPoint* and *Range* to be used are loaded. After that, state *fpu_exec* is entered, wherein the MF block awaits the FPU block to reach the desired result. As soon as it does, the FSM goes to state *fpu_result*, where the result is registered. Note that for each linguistic term, the MF block needs to compute four results as it will detailed later on in Section 6.4.2. Hence, if the computed results is not the fourth, the FSM enters state *test_step*, and iterates the process once again. Whenever all the four results are dully computed and stored, the state machine goes to state *mf_wr* where it issues the write command to MF MEM segment. The MF block iterates the same process until the last line values are written into the respective MF MEM block. Once this is done, the FSM transits to state *result* where it issues the finished signal and goes back to state *start* and waits for a new configuration stage, if any.

6.4.2 *Membership Function Memory*

As explained earlier, this memory block responds to write commands received from the MF block and read commands issued by the FU. Each word of this memory holds four data that allows the complete computation of the truth degree of a given linguistic term. The four-fold memory word contains *min*: minimum limit of the straight line; *max*: maximum limit of the straight line; *a*: angular coefficient of the straight line;*b*: linear coefficient of the straight line.

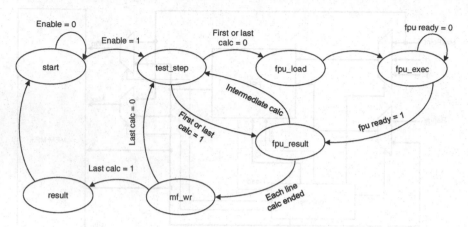

Fig. 6.5 The transition diagram of MF block controller state machine

So, every time that the MF block requests a memory write, this memory block register these values at an address, that represents the order number of the line within all the line that need to be processed, starting from zero. This block also allows the configuration of the number of lines that can be registered in the memory, which will depend on parameter Q, which determines the number of linguistic terms per membership function.

6.4.3 Fuzzification Unit

The main purpose of the fuzzification unit consists of translating the input values, returned by some sensors, to linguistic terms and respective truth degrees. This is done using the data that define the membership functions, which are stored in the MF MEM segments. Recall that for each input variable of the fuzzy system, there is a Fuzzy block associated with it.

The Fuzzy block performs the necessary computation to obtain the fuzzy version the input value. The computation consists of a comparison that may, in most cases, be followed by a multiplication then an addition, depending on the comparison result. This is repeated Q times for all the linguistic terms included in the membership function of the input variable under consideration. The Fuzzy block micro-architecture is shown in Fig. 6.6. It includes a Comparator that determines in which linguistic term range the input value falls, 2 sets of Q flip-flops to hold the result of the comparison. Their contents identify which linguistic terms are actually active. Note that more than one linguistic term may become active for the same given input variable. There are 2 sets because every linguistic term is represented by two straight lines. The Fuzzy block also includes an FPU block that is responsible for both the multiplication and addition. The obtained results for the 2 straight lines modeling the linguistic term are kept in two distinct 32-bit registers. These are the truth degrees once it is delivered by the FPU. The block includes 2 sets of 32-bit

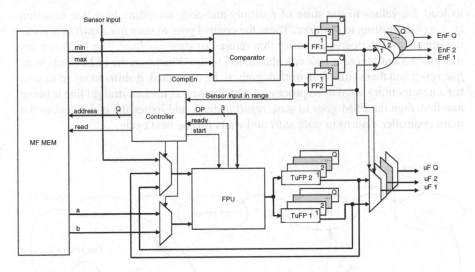

Fig. 6.6 Fuzzy block micro-architecture

registers, namely `TuFP1` and `TuFP2`, one for each linguistic term modeling the membership function of the input variable.

The inputs of a `Fuzzy` block are the characteristics of the linguistic terms of the membership function associated with the input variable under consideration. These characteristics are *a*, *b*, *min* and *max* stored in `MF MEM` segment corresponding to the input variable, as explained in Section 6.4.2. The output of a `Fuzzy` block are: signal EnF_i, for $i = 1\ldots Q$ bits, i.e. one for each included linguistic term and signal uF_i, for $i = 1\ldots Q$ 32-bit floating-point values, each of which represents the truth degree of the corresponding linguistic term. Note that linguistic terms that do not apply have 0 as a truth degree. When bit EnF_i is activated, this indicates that linguistic term number *i* of the membership function is valid with truth degree $uF_i \neq 0$. Recall that the truth degree is the product of *a* and input value augmented by *b*.

In the following, we give an overview on how the `Fuzzy` block operates. When this block receives the enable command from the main controller (see Fig. 6.2), allowing it to run, the the block controlling FSM transits from state *start* to state *mf_rd*, and triggers the `MF MEM` read command. Once the memory word is received, the FSM enters state *mf_comp*, where, using the minimum and maximum values, which represent the boundaries of the straight line associated with the linguistic term, triggers `Comparator` to perform the required comparisons to check whether the sensor input is within the boundaries of the linguistic term. Then, it shifts to state *mf_comp_result*, where it checks the result of the comparison. As every linguistic term has two lines, each result are stored in the flip-flops FF1 and FF2, as depicted in Fig. 6.7. When the comparison fails, i.e. input value is out if the prescribed range, the FSM goes directly to state *fpu_result*, and otherwise it shifts to state *fpu_load* to load the values to the suite of multiply-and-add, according to a line equation ($y = ax + b$, shifting to *fpu_exec*. Then, the control goes to state *fpu_result* to register the obtained truth degree and after that returns to state *fpu_load*. The FSM iterates this process until there is no a calculation left to do. Whenever, the FSM enters state *fpu_result* and there still some truth degrees to be computed, it shifts to *mf_rd* to wait for a new memory word to be processed. Otherwise, i.e. the last straight line is being handled, then the FSM goes to state *result* instead, and issues the end signal to the main controller, returns to state *start* and waits for the next cycle.

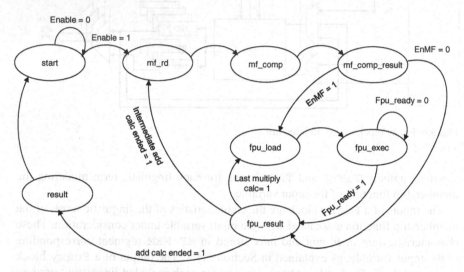

Fig. 6.7 The transition diagram of Fuzzy block controller state machine

Every EnF_i, for $i = 1 \ldots Q$ output signal is or-disjunction of the bits, registered by flip-flops FF1 and FF2. On the other hand, every truth degree uF_i, for $i = 1 \ldots Q$ is the content of one of registers TuFP 1 and TuFP 2 depending on the bit value registered in flip-flops FF1 and FF2. Note that truth degree uF_i will only be used in the subsequent inference stage if and only if $EnF_i = 1$.

6.4.4 Inference Unit

The inference unit main purpose is to identify, for each one of the output variables of the fuzzy controller, the linguistic terms that are active as well as computing the associated truth degrees. It does so using the result of the fuzzification unit and the set of predefined rules that are stored in Rules. This is the most complicated unit in the design due to the reconfigurability characteristics of the controller, as it is explained in the remaining of this section.

Before describing the details of the inference unit, let us first introduce the structure used to format the rules of the fuzzy system. A rule $\mathscr{R}$ has two defining parts:

a premise $\mathscr{P}$ and a consequent $\mathscr{C}$ as described in (6.4), wherein $\mathscr{I}_i$, for $i = 1 \dots N$ are the input variables and $\mathscr{T}_k^{\mathscr{I}_i}$ for $k = 1 \dots Q$ are the linguistic terms associated to it, $\mathscr{O}_j$, for $j = 1 \dots M$ are the output variables and $\mathscr{T}_k^{\mathscr{O}_j}$ for $\ell = 1 \dots Q$ are the linguistic terms associated with it. Note that in general the number of linguistic terms is distinct from one variable to another. However, in this work, we assume, without loss of generality, that all the variables, both of input and output, are modeled using the same number of linguistic terms Q. A rule may check only few of of the N input variables, and it may also, enable only few of the output variables.

$$\mathscr{R}: \quad \mathscr{P} \Rightarrow \mathscr{C}, \text{ where for } j, k, \ell = 0 \dots Q:$$

$$\mathscr{P} \text{ is } \mathscr{I}_0 = \mathscr{T}_j^{\mathscr{I}_0} \wedge \mathscr{I}_1 = \mathscr{T}_k^{\mathscr{I}_1} \wedge \cdots \wedge \mathscr{I}_{N-1} = \mathscr{T}_\ell^{\mathscr{I}_{N-1}} \qquad (6.4)$$

$$\mathscr{C} \text{ is } \mathscr{O}_0 = \mathscr{T}_j^{\mathscr{O}_0} \wedge \mathscr{O}_1 = \mathscr{T}_k^{\mathscr{O}_1} \wedge \cdots \wedge \mathscr{O}_{N-1} = \mathscr{T}_\ell^{\mathscr{O}_{N-1}}$$

The rule base memory `Rules` has a word size that allows to store one rule. All the rules of the model have the same structure. They include all the input output variables. When a variable is not checked or inferred, the all the linguistic terms are checked off. The binary format of a rule includes Q bits for each input and output variables, one bit for every allowed linguistic term. Hence, a rule occupies a total of $(N + M) \times Q$ bits.

The rule base memory `Rules` has P rules and thus will have $P \times (N + M) \times Q$ bits. A request to read `Rules` at some address will deliver the whole rule. The bits of the premise of are used, first of all, to check whether the rule under consideration can be fired, and if so, to trigger the computation of the truth degrees of the associated the consequent linguistic terms of the checked output variables.

A given rule fires when signal EnF_i, as delivered by the FU, for every checked of linguistic term of every input variable of the premise part of the rule under consideration is set. Furthermore, every linguistic term of any output variable that is checked in the consequent part of a fired rule need to be reported to the defuzzification unit FU. Notes that there are at most M, one for each output variable. Besides this, FU needs also to receive the truth degree for each of these checked terms.

The truth degree of an output variable linguistic term is the smallest truth degree, considering all those associated with the input variable linguistic terms in the premise part of the fired rule. When the same output variable linguistic term appears on two or more fired rules, the highest truth degree is used. Thus this done considering all the rules that fires.Recall that the truth degree of the input variable linguistic terms are provided by the FU.

Fig. 6.10 shows the micro-architecture of the `Inference` block. Its inputs consist of the Q flags EnF_i, for $i = 1 \dots Q$ and the corresponding Q truth degrees uF_i, for $i = 1 \dots Q$, which are the resulting output of FU, as described in Section 6.4.3. Its outputs are a set of M Q-bit signals EnD_i, for $i = 1 \dots M$, that identify the linguistic terms that were inferred and their respective truth degrees uD_i, for $i = 1 \dots M$, which are signals of $Q \times 32$ bits. The AND gate determines wither the current rule can be fired. The M `ANDQbits` components are simply na `AND`-arrays. In this design, the process of min-max inference is used. So, components `Minimum` and `Maximum` return the smallest of N floats and the highest of M floats, respectively. Their internal structure is omitted here for a loack of space. The `Inference` includes three memory blocks: the rule base `Rules`, a truth degree memory `MEM floats` and a bit memory `MEM bits.` given in Fig. 6.9(b) and Fig. 6.9(c), respectively. Note that initially the input is compared with 1.0 and after that with the minimum selected so far. Similarly, constant 0.0 is instead.

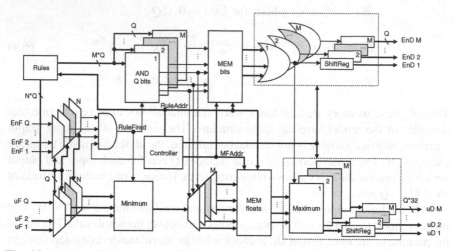

Fig. 6.8 Inference block micro-architecture

Their respective structure are shown in Fig. 6.10(a) and Fig. 6.10(b). A write request of `MEMbits` or `MEMfloats` stores a row, i.e. $M \times Q$ or $M \times Q \times 32$ bits, respectively, while a read request releases $P \times M$ and $P \times M \times 32$ bits. Assume that the ith rule of `Rules` has fired. So, the data stored in ith row of `MEMbits` consists of the consequent part of that rule, and otherwise, all the bits are reset. Similarly, the data stored in ith row of `MEMfloats` consists of the obtained minimum truth degree for that rule duplicated in all columns where a linguistic term is checked in the rule consequent.

As we can see in the Fig. 6.11, the state machine was also optimized to have the minimum number of states possible. This was done bearing in mind as to allow the reconfiguration of the number of the linguistic terms of the membership functions, rules, input output variables. Basically, there are two loops in the control imposed by the FSM. The first loop allows reading the rules one after the other and identifying the minimum of the associated truth degrees. The results of each iteration are stored in a given row of `MEMfloats`. The second loop uses the content of `MEMfloats` column after column to identify the maximize truth degree for all linguistic terms that are associated with more than one.

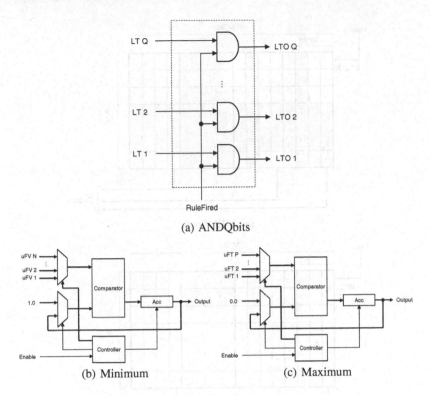

(a) ANDQbits

(b) Minimum (c) Maximum

Fig. 6.9 Internal structure of the auxiliary components: selection of the consequent part of the fired rule, the minimum and maximum truth degrees

In the following we sketch how the operation of the inference block is controlled. When the Inference block receives the enable command from the main controller allowing it to run, the state machine transits from state *start* to state *readrule*, where a specified rule to be executed is selected and read from memory Rules. Then, the control shifts to state *test_emptyrule*, where the rule loaded is checked whether it is empty. If so, the FSM goes to state *rule_result*. Otherwise, it enters state *rule_exec*. As shown in Fig. 6.10, the information of the rule premise is dispatched so as to control the operation of the two set of multiplexers. Note that there is two multiplexer for each input variable: one the flags and the other for the truth degrees. In state *rule_exec*, for each fired rule, the ANDQbits and Minimum operates simultaneously and the obtained results are stored in MEMbits and MEMfloats respectively. In the latter, before writing occurs, the results go through the set of M demultiplexers in order to associate the selected smallest truth degree to each and every one of the linguistic terms of the consequent part of the rule under consideration. The so far described process is iterated for all existing rule in the rule base. So, if the handled rule is not the last, state *readrule* is entered again and the precess is repeated. Otherwise, i.e., the last rule was processed, the FSM goes to state *mf_load* where the second loop initiates. In this state, the truth degrees of the same

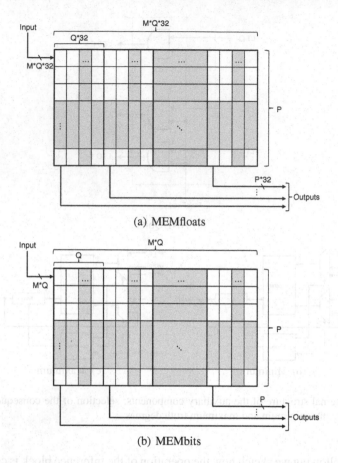

(a) MEMfloats

(b) MEMbits

Fig. 6.10 Internal structure of the auxiliary memories for inferred linguistic terms and their respective truth degrees

linguistic term of all rules are read from MEMfloats so as to provide the input to the Maximum component, which operates as soon as the FSM enters state *mf_exec*. This process is iterated Q times, which allows for the processing of the content of MEMfloats. After that, state *mf_result* is entered to shift the result in the shift register at the end of the chain in Fig. 6.11. IF there are still M columns to handle, the FSM shifts back to state *mf_load*. Otherwise, it enters to state *result*, generating the end signal to the main controller and going back to state *start* to wait for the next cycle.

6.4.5 Defuzzification Unit

The defuzzification unit main purpose is to compute the crisp value of the output variables, given the fuzzy linguistic terms and their corresponding truth values,

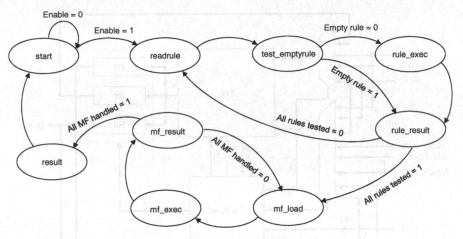

Fig. 6.11 The transition diagram of the inference controller state machine

as identified and computed by the inference unit. The centroid is used to perform the defuzzification process. Recall that uD_i for $i = 1 \ldots Q$ are the truth degrees of the linguistic terms associated with the output variable $\mathcal{O}$. This method computes the geometric center of the output membership function, considering the activated linguistic terms received from the inference block together with their respective truth degrees. The computation is done according to the steps of Algorithm 6.1.

$$\mathcal{O} = \frac{\left(\sum_{i=1}^{Q} uD_i \times mp_i \right)}{\sum_{i=1}^{Q} uD_i} \tag{6.5}$$

Algorithm 6.1. Computation of the centroid

Require: bits EnD_i and floats uD_i, $i = 1 \ldots Q$ for $\mathcal{O}$
$R_0 \Leftarrow 0; R_1 \Leftarrow 0; R_2 \Leftarrow 0$
if $EnD \neq 00 \ldots 0$ **then**
  for $i := 1$ to Q **do**
    if $EnD_i = 1$ **then**
      $R_0 \Leftarrow uD_i \times mp_i$
      $R_1 \Leftarrow R_1 + R_0$
      $R_2 \Leftarrow R_2 + uD_i$
    end if
  end for
  $R_0 \Leftarrow R_1/R_2$
end if return R_0

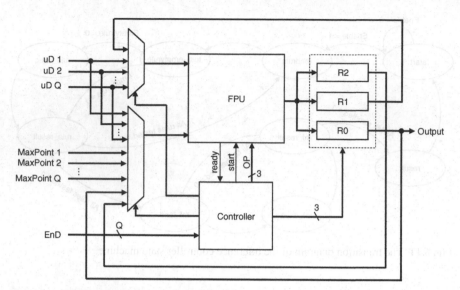

Fig. 6.12 Defuzzification block micro-architecture

Fig. 6.13 shows the state transition diagram of the FSM that controls the Defuzzy block operation. Hereafter we sketch the main steps of this control. When this block receives the enable command from the main controller allowing it to run, the FSM goes from state *start* to state *test_empty*, where the possibility of all possible linguistic terms are not enabled. If so, the FSM enters state *result*, which allows the Defuzzy to return 0 as output result. Otherwise, i.e if at least one linguistic term for the output variable that is being processed is set, in case of $EnD_i = 0$ then it goes immediately to state *fpu_result*, because this linguistic term has nothing to compute. Otherwise, in case $EnD_i = 1$, then the FSM goes to *fpu_load*, where the control enables that the values of the specific computation to be loaded and thereafter goes to state *fpu_exec*, where the computation described in Algorithm 6.1 is executed. (For the sake of clarity, the details of the necessary three operations are omitted in this description.) Once the computation performed one iteration is completed, the FSM shifts to state *fpu_result*, where it checks whether all EnD_i and respective uD_i for $i = 1 \ldots Q$ have been considered. If not, the FSM goes back to state *test_empty* and iterate repeats the same process. Otherwise, the result is readily registered and available in register R_0 and so, the FSM enters state *result*, generating the main controller's output value and the issuing a *done* signal and the block becomes ready again to operate from the start.

6.5 Performance Results

The hardware design was specified using VHDL and simulated to check its functionality using ModelSim. Subsequently, it was synthesized with the Xilinx

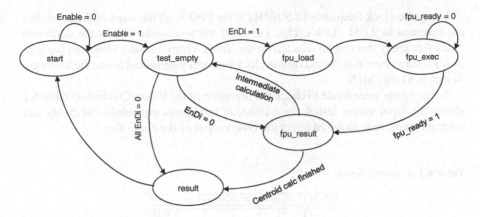

Fig. 6.13 Transition diagram for state machine that controls the defuzzification process

Plataform Studio software and implemented on the Xilinx Virtex 5 ML505-ML509-
XC5VFX70T board. The inverted pendulum application [12] is used as a benchmark. It has 2 input variables (*angle* and *velocity*, 25 rules and 1 output variable
(*speed*). The fuzzy model of the input and output variables include 5 linguistic terms
each. We can see in Fig. 6.14, the number of clock cycles that are required to execute each of the design main blocks. Note that the Defuzzy block has different
number of clock transitions, depending on how many linguistic terms were activated
as result of the Inference block. Considering the inverted pendulum, at most 3
rules can fire at a time depending on the sensors input of the Fuzzy blocks.

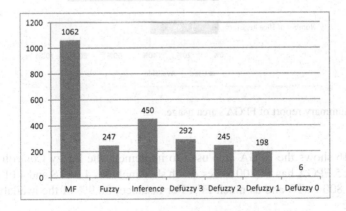

Fig. 6.14 Number of clock transitions for each block being executed by the controller at the
FPGA

Using the clock frequency of 50 MHz at the FPGA, all the steps of the controller are executed in 2,051 clock cycles, i.e. 41.02 microseconds. This is the minimum clock frequency that can be selected on the Xilinx Virtex 5 board. However, the synthesis results show that the maximum clock frequency permitted to use the hardware design is 81.407 MHz.

Several tests were made to check the precision of the Fuzzy Controller. Table 6.1 shows the input values tested, the number of fired rules the number of cycles and corresponding time required to get the crisp output of the controller.

Table 6.1 Synthesis result

Velocity	Angle	Rules fired	Cycles	Time	Speed
−1	6				0.333
−6	6	4	989	12.15	−0.667
1	−3				−0.083
−1	−5				0.500
7	−3	3	942	11.57	0.667
−11	1	2	895	10.99	−2.000
−6	−11	0	703	8.64	0.000

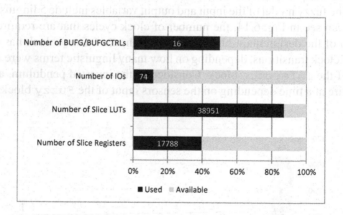

Fig. 6.15 Summary report of FPGA's area usage

Fig. 6.16 shows the FPGA area used to implement the Fuzzy Controller. Note that Virtex 5 FPGA has 11,200 Slices. Each slice includes 4 LuTS and 4 FFs, so the total is 44,800 LuTS and as much FFs. For instance, 86.9% of the available LuTs were used.

6.6 Summary

This paper proposes a massively parallel completely configurable design for fuzzy controllers. It is applicable to almost any applications in the industry that do not have

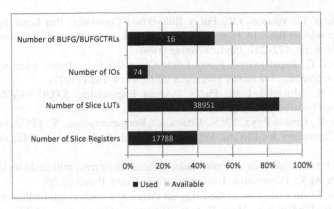

Fig. 6.16 Summary report of FPGA's area usage

a prescribed solution. The proposed architecture is parametric so that any number of inputs, outputs and rules can be accommodate with no extra effort. The design was implemented on reconfigurable FPGA and the cost and performance requirements analyzed. The next steps in the design of this controller are to investigate the generalization of the design so that to allow the use of trapezoidal and sigmoid the membership functions.

References

1. Baldwin, J.F.: Fuzzy logic and fuzzy reasoning. In: Mamdani, E.H., Gaines, B.R. (eds.) Fuzzy Reasoning and Its Applications. London Academic Press (1981)
2. Bandler, W., Kohout, L.J.: Semantics of implication operators and fuzzy relational products. In: Mamdani, E.H., Gaines, B.R. (eds.) Fuzzy Reasoning and Its Applications. London Academic Press (1981)
3. Daijin, K.: An Implementation of Fuzzy Logic Controller on the Reconfigurable FPGA System. IEEE Trans. on Industrial Electronics 47(3) (2000)
4. Diao, Y., Hellerstein, J., Parekh, S.: Using fuzzy control to maximize profits in service level management. IBM Systems Journal 41(3), 403–420 (2002)
5. Esragh, F., Mamdani, E.H.: A general approach to linguistic approximation. In: Mamdani, E.H., Gaines, B.R. (eds.) Fuzzy Reasoning and Its Applications. London Academic Press (1981)
6. Franke, K., Köppen, M., Nickolay, B.: Fuzzy image processing by using Dubois and Prade fuzzy norm. In: Proceedings of 15th International Conference on Pattern Recognition, Barcelona, Spain, pp. 518–521 (2000)
7. Fox, J.: Towards a reconciliation of fuzzy logic and standard logic. International Journal of Man-Machine Studies 15, 213–220 (1981)
8. Ghidary, S., Hattori, M., Tadokoro, S., Takamori, T.: Multi-modal human robot interaction for map generation. In: Proceedings of IEEE International Conference on Intelligent Robots and Systems, pp. 2246–2251 (2001)
9. McKenna, M., Wilamowski, B.: Implementing a Fuzzy System on a Field Programmable Gate Array Fuzzy Sets and Systems. University of Wyoming and University of Idaho

10. Magdalena, L., Velasco, J.R.: Fuzzy Rule-Based Controllers that Learn by Evolving their Knowledge Base. In: Herrera, F., Verdegay, J.L. (eds.) Genetic Algorithms and Soft Computing, pp. 172–201. Physica-Verlag (1996)
11. Mamdani, E., Pappis, C.: A Fuzzy Logic Controller for a Traffic Intersection. IEEE Trans. on Systems, Man and Cybernetics 22(6), 1414–1424 (1977)
12. Nedjah, N., Mourelle, L.M.: Fuzzy Systems Engineering. STUD FUZZ, vol. 181. Springer, Heidelberg (2005)
13. Poorani, S., Urmila Priya, T.V.S., Udaya, K., Renganarayanan, S.: FPGA based Fuzzy Logic Controllers for Electric Vehicle. Journal of the Institution of Engineers 45(5) (2005)
14. Rachel, F.M.: Proposta de um controlador automático de trens utilizando lógica nebulosa preditiva, M.Sc. Dissertation, University of São Paulo, Brazil (2006)
15. Radecki, T.: An evaluation of the fuzzy set theory approach to information retrieval. In: Trappl, R., Findler, N.V., Horn, W. (eds.) Progress in Cybernetics and System Research, Proceedings of a Symposium Organized by the Austrian Society for Cybernetic Studies, vol. 11, Hemisphere Publishing Company, NY (1982)
16. Zadeh, L.A.: Fuzzy algorithms. Information and Control 12, 94–102 (1968)
17. Zadeh, L.A.: Making computers think like people. IEEE Spectrum (8), 26–32 (1984)
18. Zadeh, L.A.: Fuzzy Logic. IEEE Computer Journal 1(83), 18 (1988)
19. Zhang, J., Knoll, A.: Designing Fuzzy Controllers by Rapid Learning. Fuzzy Sets and Systems 101, 287–301 (1999)

Chapter 7
A Reconfigurable Hardware for Subtractive Clustering*

Abstract. This chapter presents the development of a reconfigurable hardware for classification system of radioactive elements with a fast and efficient response. To achieve this goal is proposed the hardware implementation of subtractive clustering algorithm. The proposed hardware is generic, so it can be used in many problems of data classification, omnipresent in identification systems.

7.1 Introduction

Radioactive sources have radionuclides. A radionuclide is an atom with an unstable nucleus, i.e. a nucleus characterized by excess of energy, which is available to be imparted. In this process, the radionuclide undergoes radioactive decay and emits gamma rays and subatomic particles, constituting the ionizing radiation. Radionuclides may occur naturally but can also be artificially produced [1]. So, radioactivity is the spontaneous emission of energy from unstable atoms.

Correct radionuclide identification can be crucial to planning protective measures, especially in emergency situations, by defining the type of radiation source and its radiological hazard [2]. The gamma ray energy of a radionuclide is a characteristic of the atomic structure of the material.

When these emissions are collected and analyzed with a gamma ray spectroscopy system, a gamma ray energy spectrum can be produced. A detailed analysis of this spectrum is typically used to determine the identity of gamma emitters present in the source. The gamma spectrum is characteristic of the gamma-emitting radionuclides contained in the source [3].

This chapter introduces the development of a reconfigurable hardware for a classification system of radioactive elements that allow a rapid and efficient to be implemented in portable systems. our intention is to run the clustering algorithms in a portable equipment to perform the radionuclides identification. The clustering algorithms consume high processing time when implemented in software, mainly on processors of portable use, such as micro-controllers. Thus, a custom

---

* This chapter was developed in collaboration with Marcos Santana Farias.

N. Nedjah and L. de Macedo Mourelle, *Hardware for Soft Computing and Soft Computing for Hardware*, Studies in Computational Intelligence 529,
DOI: 10.1007/978-3-319-03110-1_7, © Springer International Publishing Switzerland 2014

implementation suitable for reconfigurable hardware is a good choice in embedded systems, which require real-time execution as well as low power consumption.

The rest of this chapter is organized as follows: first, in Section 7.2, is demonstrated the principles of nuclear radiation detection. Later, in Section 7.3, we review briefly existing clustering algorithms and we concentrate on the subtractive clustering algorithm. In Section 9.2, we describe the proposed architecture for cluster centers calculator using the subtractive clustering algorithm. Thereafter, in Section 9.8, we present some performance figures to assess the efficiency of the proposed implementation. Last but not least, in Section 7.6, we draw some conclusions and point out some directions for future work.

7.2 Radiation Detection

The radioactivity and ionizing radiation are not naturally perceived by the sense organs of human beings and can not be measured directly. Therefore, the detection is performed by analysis of the effects produced by radiation as it interacts with a material.

There are three main types of ionizing radiation emitted by radioactive atoms: alpha, beta and gamma. The alpha and beta are particles that have mass and are electrically charged, while the gamma rays and x-rays are electromagnetic waves. The emission of alpha and beta radiation is always accompanied by the emission of gamma radiation. So most of the detectors is to gamma radiation. Gamma energy emitted by a radionuclide is a characteristic of the atomic structure of the material. The energy is measured in electronvolts (eV). One electronvolt is an extremely small amount of energy so it is common to use kiloelectronvolts (keV) and megaelectron-volt (MeV).

Consider, for instance, Cesium-137 (Cs137) and Cobalt-60 (Co60), which are two common gamma ray sources. These radionuclides emit radiation in one or two discreet wavelengths. Cesium-137 emits 0.662 MeV gamma rays and Cobalt-60 1.33 and 1.17 MeV gamma rays. These energy are known as decay energy and define the decay scheme of the radionuclide. Each radionuclide, among many others, has a unique decay scheme by which it is identified [1].

When these emissions are collected and analyzed with a gamma ray spectroscopy system, a gamma ray energy spectrum can be produced. A detailed analysis of this spectrum is typically used to determine the identity of gamma emitters present in the source. The gamma spectrum is characteristic of the gamma-emitting radionuclides contained in the source [3].

A typical gamma-ray spectrometry system (fig. 7.1) consists of a scintillator detector device and a measure system . The interaction of radiation with the system occurs in the scintillator detector and the measurement system interprets this interaction. The scintillator detector is capable of emitting light when gamma radiation transfers to him all or part of its energy. This light is detected by a photomultiplier optically coupled to the scintillator, which provides output to an electrical signal whose amplitude is proportional to energy deposited. For gamma radiation, the most widely used scintillator is the Sodium Iodide crystal activated with thallium, NaI (Tl).

The property of these detectors provide an electrical signal proportional to the deposited energy spectrum allows the creation of the gamma energy spectrum by a radioactive element (histogram). To obtain this spectrum is used a multichannel analyzer or MCA. The MCA consists of an ADC (Analog to Digital Converter) which converts the amplitude of analog input in a number or channel. Each channel is associated with a counter that accumulates the number of pulses with a given amplitude, forming a histogram. These data form the energy spectrum of gamma radiation. As said, since different radionuclides emit radiation at different energy distributions, analyzing the spectrum can provide information on the composition of the radioactive source found and allow the identification.

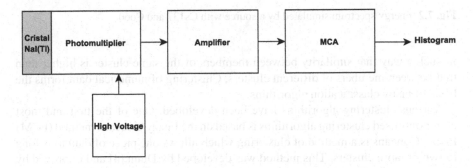

Fig. 7.1 Gama Spectrometry System - main components

Figure 7.2 shows a spectrum generated by simulation, to a radioactive source with of Cs137 and Co60. The x-axis represents the channels for a 12-bit ADC. In such a representation, 4096 channels correspond to 2.048 MeV in the energy spectrum. The first peak in channel 1324 is characteristic of Cs137 (0.662 MeV). The second and third peaks are energies of Co60.

The components and characteristics of a gamma spectrometry system (the type of detector, the time of detection , the noise of the high-voltage source, the number of channels, the stability of the ADC, temperature changes) can affect the formation of spectrum and quality of the result. For this reason it is difficult to establish a system for automatic identification of radionuclides, especially for a wide variety of these. Equipment that are in the market, using different algorithms of identification and number of radionuclides identifiable, do not have a good performance [2].

7.3 Clustering Algorithms

Clustering algorithms partition a collection of data into a certain number of clusters, groups or subsets. The aim of the clustering task is to group these data into clusters

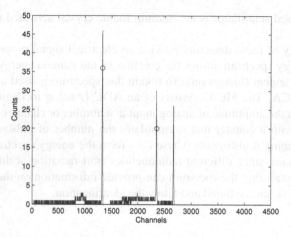

Fig. 7.2 Energy spectrum simulated by a source with Cs137 and Co60

in such a way that similarity between members of the same cluster is higher than that between members of different clusters. Clustering of numerical data forms the basis of many classification algorithms.

Various clustering algorithms have been developed. One of the first and most commonly used clustering algorithms is based on the Fuzzy C-means method (FCM). Fuzzy C-means is a method of clustering which allows one piece of data to belong to two or more clusters. This method was developed by Dunn [4] and improved by Hathaway [5]. It is commonly used in pattern recognition.

Yager and Filev [6] introduced the so-called *mountain function* as a measure of spatial density around vertices of a grid, showed in the function (7.1)

$$M(v_i) = \sum_{j=1}^{n} e^{-\alpha \|x_j - x_i\|^2}, \tag{7.1}$$

where $\alpha > 0$, M is the mountain function, calculated for the ith vertex v_i during the first step, N is the total number of data, which may be simple points or samples, that is assumed to be available before the algorithm is initiated. Norm $\| \times \|\|$ denotes the Euclidean distance between the points used as arguments and x_j is the current data point or sample. It is ensured that a vertex surrounded by many data points or samples will have a high value for this function and, conversely, a vertex with no neighboring data point or sample will have a low value for the same function. It should be noted that this is the function used only during the first step with all the set of available data. During the subsequent steps, the function is defined by subtracting a value proportional to the peak value of the mountain function. A very similar approach is the subtractive clustering (SC) proposed in [7]. It uses the so-called *potential* value defined as in (7.2).

$$P_i = \sum_{j=1}^{n} e^{-\alpha \|x_j - x_i\|^2}, \text{ where } \alpha = \frac{4}{r_a} \tag{7.2}$$

wherein, P_i is the potential-value i-data as a cluster center, x_i the data point and r_a a positive constant, called *cluster radius*.

The potential value associated with each data depends on its distance to all its neighborhoods. Considering (7.2), a data point or sample that has many points or samples in its neighborhood will have a high value of potential, while a remote data point or sample will have a low value of potential. After calculating potential for each point or sample, the one, say x_i^*, with the highest potential value, say P_i^*, will be selected as the first cluster center. Then the potential of each point is reduced as defined in (7.3). This is to avoid closely spaced clusters. Until the stopping criteria is satisfied, the algorithm continues selecting centers and revising potentials iteratively.

$$P_i = P_i - P_i^* e^{-\beta \|x_i - x_i^*\|^2}, \tag{7.3}$$

In (7.3), $\beta = 4/r_b^2$ represents the radius of the neighborhood for which significant potential revision will occur. The data points or samples, that are near the first cluster center, say x_i^*, will have a significantly reduced density measures. Thereby, making the points or samples unlikely to be selected as the next cluster center.

The subtractive clustering algorithm can be briefly described by the following 4 main steps:

- Step 1: Using (7.2), compute the potential P_i for each point or sample, $1 \le i \le n$;
- Step 2: Select the data point or sample, x_i^*, considering the highest potential value, P_i^*;
- Step 3: Revise the potential value of each data point or sample, according to (7.3);
- Step 4: If $maxP_i \le \varepsilon P_i^*$, wherein ε is the reject ratio, terminate the algorithm computation; otherwise, find the next data point or sample that has the highest potential value and return to Step 3.

The main advantage of this method is that the number of clusters or groups is not predefined, as it is in the fuzzy C-means method, for instance. Therefore, this method becomes suitable for applications where one does not know or does not want to assign an expected number of clusters á priori. The cluster estimates obtained by the subtractive clustering can be used to initialize iterative optimization-based clustering methods and as well as the set of rules used in fuzzy clustering methods.

7.4 Proposed Architecture

This section provides an overview of the macro-architecture and contains information on the broad objectives of the proposed hardware. The hardware implements the subtractive clustering algorithm. The subtractive clustering algorithm was briefly explained in the previous section.

The implementation of this algorithm in hardware is the main point is to develop a classification system of radioactive elements. For referencing, this hardware it will call HSC, hardware to subtractive clustering. This hardware processes all the arithmetic computation, described in the section above, to calculate the potential of each point in the subtractive clustering algorithm.

The other component of this macro-architecture will be called SLC, component to storage, loading and control, which provides to the HSC the set of samples for the selection of cluster centers and stores the results of the calculated potential of each sample. This component also has the controller of the HSC. Figure 7.3 shows the components of the described macro-architecture.

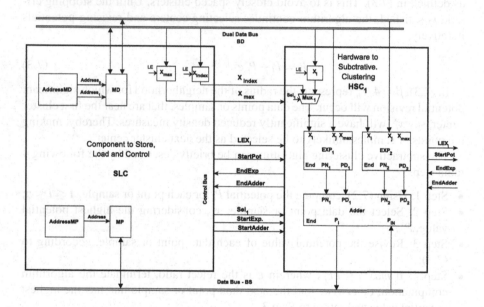

Fig. 7.3 Macro-architecture components - SLC e HSC

The SLC is a controller based on state machine. It includes a dual port memory MD that provides the data that has to be clustered and memory MP that allows for the bookkeeping of the potential associated with each clustered data. The registers X_{max}, X_i and X_{Index} maintain the required data until component EXP$_1$ and EXP$_2$ have completed the related computation. We assume the X_{max} value is available in memory MD at address 0. The X_{max} is the biggest value found within the data stored in MD. This register is used to the data normalization.

The two EXP components, inside HSC, receive, at the same time, different x_j values from the dual port memory MD. So the two modules start at the same time and thus, run in parallel. After the computation of $e^{-\alpha\|x_i-x_j\|^2}$ by EXP_1 and EXP_2, component ADDER sums and accumulates the values provided at its input ports. This process is repeated until all data x_j, $1 \leq j \leq N$, are handled. So, this computation yileds the first P_i value to be stored in memory MP. After that, the process is repeated to compute the potential values of all data points in memory MD. At this point, the first cluster center has been found.

The SLC component works as a main controller of the process. Thus, the trigger for initiating the processing components EXP_1 and EXP_2 occurs from the signal *StartExp* sent by SLC. The component SLC has a dual-port memory MD which stores the samples / points to be processed. Memory MD allows the two components (EXP_1 and EXP_2) receiving a sample to calculate the exponential value and thus can operate in parallel . This sample for each component EXP are two distinct values x_j from two subsequent memory addresses.

The proposed architecture allows the hardware to subtractive clustering HSC can be scaled by adding more of these components in parallel to the computation of the factors $e^{-\alpha\|x_j-x_i\|^2}$. This provides greater flexibility to implement the hardware. Figure 7.4 shows how new components HSC are assembled in parallel.

Each component HSC calculates in parallel the potential of a point i, the value P_i of the function 7.3. For this reason each module (HSC) must to receive and record a value of x_i to work during the calculation of the potential of a point. Since these values are in different adrress of the memory, this registry value x_i has to be done at different time because the memory can not have your number of ports increased as the number of components HSC is increased. To be not necessary to increase the number of control signals provided by the component SLC when new components HSC are added, the component HSC itself has to send some control signals for the thereafter.

These signs are to load the value x_i (*LEX$_i$*) and start the reduction potential of each point (*StartPot*), as showed in 7.3. Moreover, each component HSC should receive the signal *EndAdd* which indicates the end of the operation on the component

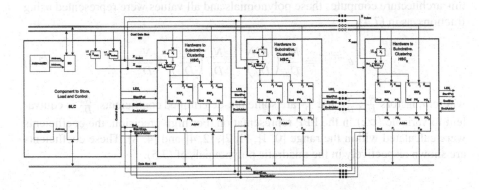

Fig. 7.4 Macro-architecture with HSC components in parallel

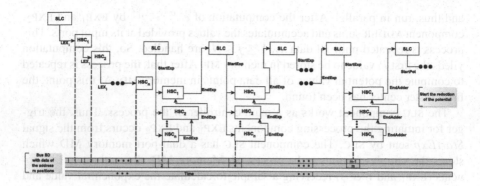

Fig. 7.5 Control signals with scaled architecture

ADDER of the thereafter component HSC. This ensures that the main control (SLC) only receive these signals after all the components of the HSC in parallel complete their transactions at each stage, allowing the hardware can be reconfigured without change in the main control. Figure 7.5 shows the effect of this scaling, simulating different processing times between the HSC.

The n components HSC, implemented in parallel, compute the potential of n points of the set of samples. As explained earlier, the record value of x_i, to be used in the calculation of the potential it has to be done in time different. It is shown in figure 7.5 that the first component HSC receives the signal LEX_i from SLC control and after registering it x_i, it sends the signal LEX_i for HSC thereafter. Only after all of the HSC to have recorded its value x_i, the signal to start the components EXP ($StartExp$) is sent with the first pair of values x_j in the dual bus BD.

Fig. 7.7 shows the architecture of the module EXP_1 and EXP_2 that permits the calculation of the exponential value $e^{-\alpha\|x_i - x_j\|^2}$. The exponential value was approximated by a second-order polynomial using the least-squares method [8]. Moreover, this architecture computes these polynomials and all values were represented using fractions, as in (7.4).

$$e^{-\alpha\|x\|} = \frac{N_a}{D_a}\left(\frac{N_v}{D_v}\right)^2 + \frac{N_b}{D_b}\left(\frac{N_v}{D_v}\right) + \frac{N_c}{D_c} \qquad (7.4)$$

wherein, factors $\frac{N_a}{D_a}$, $\frac{N_b}{D_b}$ and $\frac{N_c}{D_c}$ are some pre-determined coefficients. $\frac{N_v}{D_v}$ is equivalent to variable (αx) in the FPP representation. For high precision, the coefficients were calculated within the range [0, 1[, [1, 2[, [2, 4[and [4, 8[. These coefficients are shown respectively in the quadratic polynomials of (7.5).

$$e^{-(\alpha x)} \cong \begin{cases} P_{[0,1[}(\frac{N_v}{D_v}) = \frac{773}{2500} \left(\frac{N_v}{D_v}\right)^2 - \frac{372}{400} \left(\frac{N_v}{D_v}\right) + \frac{9953}{10000} \\[2mm] P_{[1,2[}(\frac{N_v}{D_v}) = \frac{569}{5000} \left(\frac{N_v}{D_v}\right)^2 - \frac{2853}{5000} \left(\frac{N_v}{D_v}\right) + \frac{823}{1000} \\[2mm] P_{[2,4[}(\frac{N_v}{D_v}) = \frac{67}{2500} \left(\frac{N_v}{D_v}\right)^2 - \frac{2161}{10000} \left(\frac{N_v}{D_v}\right) + \frac{4565}{10000} \\[2mm] P_{[4,8[}(\frac{N_v}{D_v}) = \frac{16}{10000} \left(\frac{N_v}{D_v}\right)^2 - \frac{234}{10000} \left(\frac{N_v}{D_v}\right) + \frac{835}{10000} \\[2mm] P_{[8,\infty[}(\frac{N_v}{D_v}) = 0 \end{cases} \quad (7.5)$$

The accuracy of these calculated values, i.e. the introduced error is no more 0.005, is adequate to properly obtain the potential values among the data provided during the process of subtractive clustering. The absolute error introduced is shown in Fig. 7.6. Depending on the data, this requires that the number of bits to represent the numerator and denominator have to be at least twice the maximum found in the data points provided.

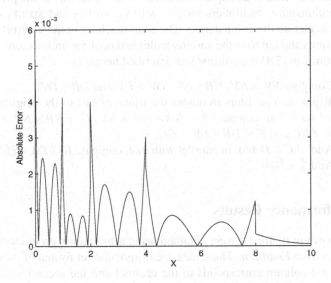

Fig. 7.6 Absolute error introduced by the approximation computation

The architecture of the Fig. 7.7 presents the micro-architecture of components EXP$_1$ and EXP$_1$. It uses four multipliers, one adder/subtracter and some registers. These registers are all right-shifters. The controller makes the adjustment of the binary numbers with shifts to the right in these registers in order to maintain the frame

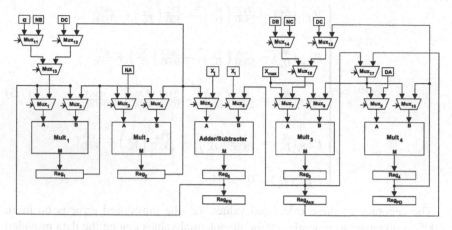

Fig. 7.7 Architecture of EXP Modules to compute the exponential $e^{-\alpha\|x_i-x_j\|^2}$

of binary numbers after each operation. This is necessary to keep the results of multiplication with the frame of bits used without much loss of precision. The closest fraction is used instead of a simple truncation of the higher bits of the product.

In this architecture, multipliers $MULT_1$, $MULT_2$, $MULT_3$ and $MULT_4$ operate in parallel to accelerate the computation. The state machine in the controller triggers these operations and controls the various multiplexers of the architecture. The computation defined in (7.4) is performed as described hereafter.

- Step 1: Compute $NV \times NV$, $NB \times NV$, $DV \times DV$ and $DB \times DV$;
- Step 2: Right-shift registers to render the frame of bits to the original size and in parallel with that, compute $A = NA \times NV \times NV$, $C = NB \times NV \times DC$, $D = DB \times DV \times NC$ and $E = DB \times DV \times DC$;
- Step 3: Add of $C + D$ and, in parallel with that, compute $B = DA \times DV \times DV$;
- Step 4: Add $\frac{A}{B} + \frac{C+D}{E}$.

7.5 Performance Results

The data shown in figure 7.2 were obtained using a simulation program called *Real Gamma-Spectrum Emulator*. These data are in spreadsheet format of two columns, where the first column corresponds to the channel and the second is the number of counts accumulated in each channel. To validate the method chosen (subtractive clustering), the algorithm was implemented with Matlab, using the simulated data. As seen in the introduction, these data simulate a radioactive source consists of Cs137 and Co60. To apply the subtractive clustering algorithm in Matlab data provided by the simulation program needed to be converted into one-dimensional data in one column. For example, if channel 1324 to accumulate 100 counts, means that the value 1324 should appear 100 times as input. only in this way the clustering algorithm is able to split the data into subgroups by frequency of appearance. In a real application this data would be equivalent to the output of AD converter of a gamma spectrometry system, as shown in the introduction.

In the spectrum of Fig. 7.2, one can see three peaks. The first one in the channel 1324 is characteristic of Cs137 (0.662 MeV). The second and third peaks correspond the energy of Co60. The circular marks near the first and second peaks show the result of applying the subtractive clustering algorithm on the available data with Matlab software. These circular marks are center of the found clusters. These found clusters are very near (one channel to the left) of the signal peaks, the expected result. With the configuration to the algorithm in Matlab, the third peak was not found. This result can change with an adjust of the radius r_a in 7.2. This is enough to conclude that the data provided belongs to a radioactive source with Cs137 and Co60 and the subtractive cluster method can be used to identify these radionuclides.

As the proposed architecture is based on the same algorithm, is expected to find the same result. The initial results show that the expected cluster center can be identified as in Matlab specification. The hardware takes about 12660 clock cycles to yield one sum of exponential values ($\sum_{j=1}^{n} e^{-\alpha\|x_i-x_j\|^2}$). Considering the one hundred points in the avaiable data set of the case study, the identification of the first cluster center would take ten times that amount, i.e. about 126600 clock cycles. However, finding the center of the second cluster is faster. It should take about 13000 clock cycles. This result can change with the data and depends of the amount of adjustment required to the right in the shift registers during the process. The simulation results of an instance of this process is shown in Fig. 7.8.

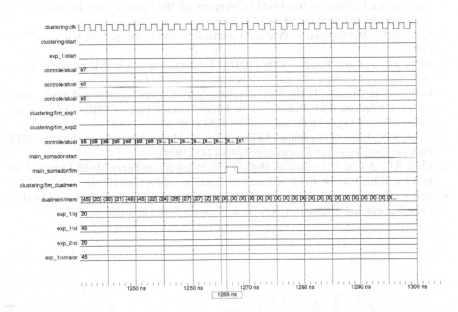

Fig. 7.8 Simulao de forma de onda de deslocamentos a direita para ajuste no nmero de bits do resultado

7.6 Summary

This chapter describes the implementation of subtractive clustering algorithm in hardware. The results shows the expected cluster center can be identified with a good efficiency. In data from the simulation of signals of radioactive sources, after conformation of the signal and its conversion into digital , the cluster center represents the points that characterize the energy provided by a simulated radionuclides. The identification of these points can sort the radioactive elements present in a sample. With this hardware it was possible to identify more than one cluster center, which would recognize more than one radionuclide in radioactive sources.

These results reveal that the proposed hardware can be used to develop a portable system for radionuclides identification. This system can be developed and enhanced integrating the proposed hardware with a software to be executed by a processor inside the FPGA, bringing reliability and faster identification, an important characteristics for these systems. Following this work, we intend to develop a software-only implementation using an embedded processor or a micro-controller to compare it with the hardware-only solution.

References

1. Knoll, G.F.: Radiation Detection and Measurement. John Wiley and Sons, New York (1989)
2. Performance Criteria for Hand-held Instruments for the Detection and Identification of Radionuclides. ANSI Standard N42.34 (2003)
3. Gilmore, G., Hemingway, J.: Practical Gamma Ray Spectrometry. John Wiley and Sons (1995)
4. Dunn, J.C.: A Fuzzy Relative of the ISODATA Process and Its Use in Detecting Compact Well-Separated Clusters. Journal of Cybernetics 3, 32–57 (1973)
5. Hathaway, R., Bezdek, J., Hu, Y.: Generalized fuzzy C-means clustering strategies using Lp norm distances. IEEE Transactions on Fuzzy Systems, Proc. of SPIE Conf. on Application of Fuzzy Logic Technology, pp. 246–254 (1993)
6. Yager, R.R., Filev, D.: Learning of Fuzzy Rules by Mountain-Clustering. In: Proc. IEEE Internat. Conf. on Fuzzy Systems, pp. 1240–1245 (1994)
7. Chiu, S.L.: A Cluster Estimation Method with Extension to Fuzzy Model Identification. In: Proc. IEEE Internat. Conf. on Fuzzy Systems, pp. 1240–1245 (1994)
8. Rao, C., Toutenburg, H., Fieger, A., Heumann, C., Nittner, T., Scheid, S.: Linear Models: Least Squares and Alternatives, New York. Springer Series in Statistics (1999)

Chapter 8
Reconfigurable Hardware for DNA Matching*

Abstract. DNA sequence matching is used in the identification of a relationship between a fragment of DNA and its owner by mean of a database of DNA registers. A DNA fragment could be a hair sample left at a crime scene by a suspect or provided by a person for a paternity exam. The process of aligning and matching DNA sequences is a computationally demanding process. In this chapter, we propose a novel parallel hardware architecture for DNA matching based on the steps of the BLAST algorithm. The design is scalable so that its structure can be adjusted depending on size of the subject and query DNA sequences. Moreover, the number of units used to perform in parallel can also be scaled depending some characteristics of the algorithm. The design was synthesized and programmed into FPGA. The trade-off between cost and performance were analyzed to evaluate different design configuration.

8.1 Introduction

Bioinformatics is a field of biological science which deals with the study of methods for storing, retrieving and analyzing biological data such as DNA. It also involves finding the genes in the DNA sequences of various organisms, developing methods to predict the structure and/or function of newly discovered proteins and structural RNA sequences, clustering protein sequences into related families. Specifically, it includes solving the problem of aligning similar proteins in general and DNA in particular 10.

One of the main challenges in bioinformatics consists of aligning DNA. DNA stripes are long sequences of DNA bases, which are represented as A (Adenine), C (Cytosine), G (Guanine) and T (Thymine). In this sense, algorithms are specifically developed to reduce time spent in DNA alignment and matching, evaluating similarity degree between the subject and the query sequence. These algorithms are usually based on dynamic programming, which work well providing a fair tradeoff

---

* This chapter was developed in collaboration with Edgar José Garcia Neto Segundo.

N. Nedjah and L. de Macedo Mourelle, *Hardware for Soft Computing and Soft Computing for Hardware*, Studies in Computational Intelligence 529,
DOI: 10.1007/978-3-319-03110-1_8, © Springer International Publishing Switzerland 2014

between time and cost for short sequences. However, commonly these algorithm take exponentially more time as DNA sequences get longer.

The major advantage of the methods based on dynamic programming are the commitment to discover the best match. However, that commitment requires huge computational resources [7, 4]. DNA matching algorithms based on heuristics [8] emerged as an alternative to dynamic programming in order to reduce the required high computational cost. Instead of aiming at the best alignment(s), heuristics-based methods attempt to find a set of acceptable or pseudo-optimal matches. Ignoring unlikely alignments, these techniques have improved the performance of DNA matching [3, 5, 10]. Among heuristics-based methods, BLAST [1, 2] and FASTA [9, 7] stand out. Both of these algorithms have well defined procedures for the three main stages of aligning algorithms, which are seeding, extending and evaluating. BLAST is the fastest algorithm known so far [1, 2, 6]. In this chapter, we focus of this algorithm and propose a massively parallel architecture suited as an ASIC for DNA matching using BLAST. The main objective of this work is the acceleration of the aligning and matching procedures.

This chapter is organized as follows: First, in Section 8.2, we sketch briefly the steps used in the BLAST algorithm; Thereafter, in Section 9.2, we detail the proposed parallel architecture, pointing out specifically its scalability characteristics; Subsequently, in Section 9.8, we describe the setup used to implement the proposed architecture on FPGAs and evaluate the performance of the design; Finally, in Section 8.5, we draw some concluding remarks and point out directions for future work.

8.2 BLAST Algorithm

The BLAST (Basic Local Alignment Search Tool) [1] algorithm is a heuristic search-based method that seeks words in the subject sequence s of length w that score at least T, called the *alignment threshold*, when aligned with the query sequence t. The scoring process is performed according to predefined criteria that are usually prescribed by geneticists. This task is called *seeding*, where BLAST attempts to find regions of similarity to begin its matching procedure. This step has a very powerful heuristic advantage, because it only keeps pairs whose matching score is larger than the pre-defined threshold T. Of course, there is some risk of leaving out some worthy alignments. Nonetheless, using this strategy, the search space decreases drastically, and hence accelerating the convergence of the matching process.

After identifying all possible alignments locations or *seeds*, the algorithm proceeds with the *extension stage*. It consists of extending the found alignments to the right and left within both the subject and query sequences, in an attempt to find a locally optimal alignment. Some versions of BLAST introduce the use of a wild-card symbol (_), called the *gap*, which can be used to replace any mismatch [7, 10]. Here, we do not allow gaps. Finally, BLAST try to improve score of high scoring pairs, HSP, through a second extension process and the dismissal of a pair is done when the corresponding score does not reach a new pre-defined threshold. HSPs that meet this criterion will be reported by BLAST as final results, provided that they do not exceed the cutoff prescribed value, which specifies for number of descriptions and/or alignments that should be reported. This last step is called *evaluation*. In the implementation presented in this chapter, we do not assess the results provided by the extension stage. We simply provide all of them as a final result of the alignment process.

BLAST employs a measure based on a well-defined mutation scores. It directly approximates the results that would be obtained by any dynamic programming algorithm for optimizing this measure. The method allows for the detection of weak but biologically significant similarities. The algorithm is more than one order of magnitude faster than existing heuristic algorithms. Compared to other heuristics-based methods, such as FASTA [7], BLAST performs DNA and protein sequence similarity alignment much faster but it is considered to be equally sensitive.

The BLAST algorithm proceeds through three main steps: *(i)* seeding, which allows to find and mark all seeds. These are subsequences of size w that can be considered as alignment points. Algorithm 8.4 describe the work as it should be done during this step; *(ii)* extension, which extends at most, i.e. with respect to the limits of the subject and query sequences, all the marked seeds and marks all those extensions that scored more that the prescribed threshold T. The extension is done in both directions, i.e. to the right of the seed location in the subject and query sequences as well as to the its left; Algorithm 8.2 describes the extension done to the right of the seed. Note that the algorithm does the extension to the left (Algorithm 8.4) is similar to the one presented with the exception that sequence counters i and j are decremented and the base are appended to the left; *(iii)* assessment, which selects some of the alignments, as found by the extension stage, and applies some biological parameters to extract some few promising alignment to be considered further in the DNA matching biological process. This last step, as described in Algorithm 8.3, is not treated any further in this chapter.

Algorithm 8.1. Seeding procedure

Require: Subject and query sequences s and t respectively
Ensure: Matrix of seed location *hits*
1: **let** $s = [s_0, s_1, \ldots, s_i, \ldots, s_{m-1}]$
2: **let** $t = [t_0, t_1, \ldots, t_j, \ldots, t_{n-1}]$
3: $sws \leftarrow [sw_0, sw_1, \ldots, sw_i, \ldots, sw_{m-w}]$, wherein $sw_i = [s_i, s_{i+1}, \ldots, s_{i+w-1}]$
4: $tws \leftarrow [tw_0, tw_1, \ldots, tw_j, \ldots, tw_{n-w}]$, wherein $tw_j = [t_j, t_{j+1}, \ldots, t_{j+w-1}]$
5: **for** $i = 0 \rightarrow (m - w)$ **do**
6:      **for** $j = 0 \rightarrow (n - w)$ **do**
7:          **if** $tw_i = sw_j$ **then**
8:              $hits[i, j] \leftarrow 1$
9:          **else**
10:            $hits[i, j] \leftarrow 0$
11:         **end if**
12:      **end for**
13: **end for**

Algorithm 8.2. Extension procedure (right)

Require: Sequences s and t, seed offsets i and j respectively
Ensure: Extension score σ

1: $E_s \leftarrow sw_i \boxplus s_{i+w}; k \leftarrow i$
2: $E_t \leftarrow tw_j \boxplus t_{j+w}; \ell \leftarrow j$
3: **repeat**
4: $\quad k \leftarrow k+1; E_s' \leftarrow E_s; E_s \leftarrow E_s \boxplus s_{k+w+1}$
5: $\quad \ell \leftarrow \ell+1; E_t' \leftarrow E_t; E_t \leftarrow E_t \boxplus t_{\ell+w+1}$
6: **until** $(s_{k+w+1} \neq t_{\ell+w+1})$ or $(k > m-1)$ or $(\ell > n-1)$
7: **if** $s_{k+w+1} = t_{\ell+w+1}$ **then**
8: $\quad \sigma \leftarrow Scores(E_s, E_t)$
9: **else**
10: $\quad \sigma \leftarrow Scores(E_s', E_t')$
11: **end if**

Algorithm 8.3. Assessment procedure

Require: Offsets i, j, threshold T and extension score σ
Ensure: Matrix $hits$ updated

1: **if** $\sigma \geq T$ **then**
2: $\quad hits[i, j] \leftarrow \sigma$
3: **else**
4: $\quad hits[i, j] \leftarrow 0$
5: **end if**

Algorithm 8.4. Extension procedure to the let

Require: s, t and $hits$ as a results of seeding;
Ensure: $hits$ updated

1: $E_s \leftarrow s_{i+w} \boxplus sw_i; k \leftarrow i$
2: $E_t \leftarrow t_{j+w} \boxplus tw_j; \ell \leftarrow j$
3: **repeat**
4: $\quad k \leftarrow k-1; E_s' \leftarrow E_s; E_s \leftarrow s_{k+w+1} \boxplus E_s$
5: $\quad \ell \leftarrow \ell-1; E_t' \leftarrow E_t; E_t \leftarrow t_{\ell+w+1} \boxplus E_t$
6: **until** $(s_{k+w+1} \neq t_{\ell+w+1})$ or $(k < 0)$ or $(\ell < 0)$
7: **if** $s_{k+w+1} = t_{\ell+w+1}$ **then**
8: $\quad \sigma \leftarrow Scores(E_s, E_t)$
9: **else**
10: $\quad \sigma \leftarrow Scores(E_s', E_t')$
11: **end if**

8.3 Proposed Architecture

The overview of the proposed architecture is depicted in Fig. 8.1. The Hardware HBLAST implements the BLAST algorithm, as described in Section 8.2. Besides the clock signal, it receives as input the subject and query sequences of m and n

bases respectively. Note that, in general, we have $m \ll n$. HBLAST also expects the configuration of three parameters: w, which determine the seed size, T, which sets up the required threshold value for alignment acceptance during extension, and p, which dictates the number of extension processor that will be used in parallel as it will be show later.

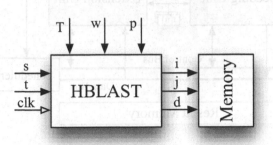

Fig. 8.1 Interface of the proposed design

As there are 4 DNA bases (A, C, G, and T), we need 2 bits to represent each base distinctively (00, 01, 10, 11). Instead of representing the subject and query sequences 2 registers of $2 \times m$ and $2 \times m$ bits respectively, we opted to use 2 registers of m bits to store subject sequence and 2 registers of n bits to hold the query sequence: one register of the pair holds the MSB of the DNA bases that form the sequence and the other the LSB. These two ways of storing the DNA sequences require the same number of flip-flops, but the second way improves the matching time as the two bits of a base can be compared in parallel without much increase in control, as they are provide by two distinct registers. The macro-architecture of HBLAST is given in Fig. 8.2. It includes a Seeding Unit that takes care of finding and bookkeeping all the seeds, with respect to s and t, and an Extension Unit that extends the seeds found.

A Global Controller synchronizes the work in pipeline of the seeding and extension units: seeds are handled by the Extension Unit as they come. There is no need to complete the seeding step before starting the extension work. A Scheduler arbitrates the use of the shared data and control buses between the Seeding and Extension Units. This is necessary because the Seeding Unit is, in turn, structurally formed by $q = n - w + 1$ concurrent sub-units and the Extension Unit is formed by p extension processors that act in parallel to accelerate the alignment process. The structural parallelism within the Seeding and Extension Unit is depicted in Fig. 8.3. The work of the q seeding components (Seeding$_i$) and the p extension components (Extension Processor$_j$) is harmonized by a respective stage controller, i.e. the Seeding Controller and the Extension Controller respectively.

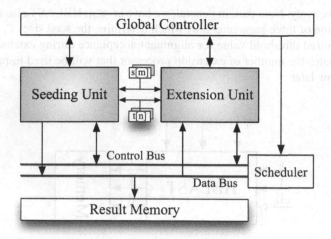

Fig. 8.2 Proposed macro-architecture

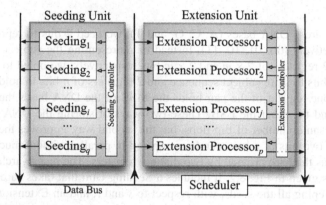

Fig. 8.3 Structural parallelism in the seeding and extension units

8.3.1 Seeding Unit

The design uses $q = n - w + 1$ concurrent Seeding components. Fig. 8.4 describes
the corresponding micro-architecture along with the interface withe the Scheduler
and the Seeding Controller. Each of these Seeding components includes 2 Matching
Units: one for the comparison of the MSBs of subject and query DNA sequences
and the other for the LSBs. The Matching Unit is a mere array of w XNOR gates
whose results are summarized by an AND gate, as shown in the circuit of Fig. 8.5(a).

When a match of a target (w consecutive bits of s) and a word (w consecutive bits of t) is declared, i.e. the result of the both Matching units (MSB and LSB) are both 1, the stamp formed by the offset of the target and word is pushed down the FIFO. Note that there is one FIFO per Seeding Unit. The stamps are later popped to be considered for extension. Once a FIFO (or a Seeding Unit) is selected by the Scheduler to feed a requesting Extension Processor, the Write Logic of Fig. 8.5(b) allows the output stamp of the FIFO to be written into the Data Bus so as to be forwarded to the Extension Processor.

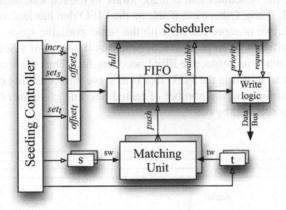

Fig. 8.4 Seeding unit micro-architecture

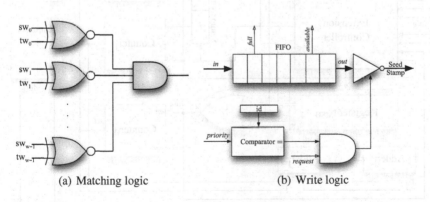

(a) Matching logic   (b) Write logic

Fig. 8.5 Matching and write logic micro-architectures

8.3.2 *Extension Unit*

The Extension Unit includes p Extension Processors as shown in Fig. 8.6. The number of included processor is defined as external parameter. This number does not necessarily coincide with that of Seeding components as many seeds do not require much extension work. Some seeds are discarded in the first base extension. Note that it is intended that $p \ll q$. For this purpose, among others, a Scheduler is used to distributed the identified seeds (in the FIFOs) as soon as a processors becomes idle. When a processor completes the extension of a given seed and requests a new one to work with, the Scheduler that is made aware of the request, selects the FIFO that is already full, if any. Otherwise, it selects the FIFO that has less available space. In the case there two or more FIFOS with the same available space, the one with the smallest identifier is given precedence. Note that the work of a Seeding component is suspended when its respective FIFO becomes full. Thus the strategy adopted by the Scheduler in selecting the FIFO that is to serve the requesting extension processor aims at minimizing the number of halted Seeding components. As soon as an interruption is received by the

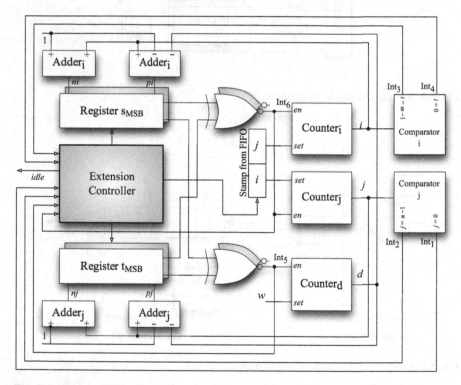

Fig. 8.6 Micro-architecture of the Extension Processor

Extension to the right and left are done parallel. The extension processor includes 4 adders that compute the new offset as well as the length of the matched subsequence. During the extension to the right, 2 adders compute $ni \leftarrow i + 1$ and $nj \leftarrow j + 1$ while during the extension to the left, the other 2 adders compute $pi \leftarrow i - d - 1$ and $pj \leftarrow j - d - 1$. These new indices allow the processor to have access to the new bases at the immediate left and right to the d bases already matched. At first, we have $d = w$, then d is incremented at every successful match. The actual update of indices i, j and d is done, by the counters, only once the match is declared. When a mismatch occurs, an interrupt (Int_6 or Int_5) is triggered to abandon the current seed. Two other interrupts can occur when the either all bases to seed's right or left on t (Int_1 or Int_2) or t those to the seed's right or left on s were treated. When interruption occurs, the Extension Controllers enables the writing of the triplet (i, j, d) into the Result Memory and signals to the Global Controller that the Extension Processor in question is idle and thus generates s request for a new seed to work pass it through to the processor.

8.3.3 The Controllers

The design includes 4 controllers: the Global Controller, the Seeding Controller and the Extension Controller and the Scheduler. Controllers are implemented as finite state machines.

The Global Controller is responsible mainly for the synchronization of the pipeline between the seeding and extension stages. Besides, it allows for the initialization of all components, the load of the DNA subject and query sequences into the corresponding registers and enabling the writing operation of the final results into the Result Memory.

The actions imposed by the seeding Controller guarantee the logic distribution of the DNA sequences into targets words son as to allow for the matching process to perform correctly. The main task of this controller consist of maintaining the content of register s and t coherent all the time by synchronizing the required shifting operations.

The Extension Controller is responsible for the correct performance of the p Extension Processors. It handles the interruption signals send by the Extension Processors and controls the injection of the bits that represent the bases that need to be considered during extension to the right and/or left, depending on the status of the triggered interruptions.

The Scheduler is responsible for controlling the use of the Data Bus as to forward an give seed stamp to an identified Extension Processor. It also selects the FIFO that needs to provide the seed stamp to be treated next when the Extension Controller signals that one of the Extension Processor became idle.

8.4   Performance Results

The MicroBlaze[TM] and the co-processor HBLAST were synthesized in a Xilinx Virtex 5 FPGA xc5vfx70t. The MicroBlaze is an embedded processor soft core, which is a reduced instruction set computer optimized for implementation on Xilinx[TM] FPGAs.

Without the proposed HBLAST co-processor, the MicroBlaze processor performs all the alignment process. In this case, the BLAST algorithm were implemented in ANSI/C++. The MicroBlaze has a communication interface for point-to-point, called Fast Simplex Link (FSL), which allows for an efficient connection with an external co-processor. In the remainder of this section, we will first introduce the performance figures of the HBLAST proposed design in terms of area and time requirements, then we compare the performance of the Microblaze-based implementation (software implementation) and that occasioned by the use of HBLAST as a co-processor (hardware implementation).

Table 8.1 shows the impact of varying the number of bases in the subject and query sequences on both area and time requirements. Note that in case 4, wherein $m = 100$ and $n = 25$, the hardware resources available on the used FPGA were exhausted and thus no time figure is given in this case. Fig. 8.7 illustrates graphically this impact.

Table 8.1 Hardware area and time requirements for diffrent configuration of m and n

#	m	n	FFs	%	LuTs	%	Slices	%	Time
1	20	10	7887	18	7811	17	3300	29	12.59
2	60	20	33418	75	33124	74	10900	97	14.91
3	100	10	27767	62	28307	63	9547	85	19.57
4	100	25	49907	111	50952	114	12411	111	—

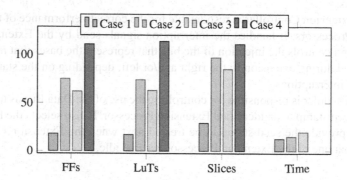

Fig. 8.7 Impact of the number of seed bases on the area and time requirements

Table 8.2 shows the impact of the value chosen for the seed size w. It is possible to note that adjusting the setting of this parameter can be a way to remedy to the case when the hardware are required is slightly above the available resources. Note that in this case, we set $m = 20$, $n = 10$ and $p = 2$. A graphical illustration of this effect is shown in Fig. 8.8.

Table 8.2 Hardware area and time requirements as w increases

w	FFs	%	LuTs	%	Slices	%	Time
3	7887	18	7811	17	3300	29	12.59
4	7205	16	7278	16	3258	29	7.64
5	6523	15	6689	15	2963	26	5.71
6	5841	13	4553	10	2631	23	4.86

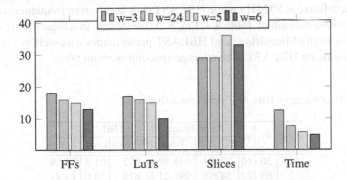

Fig. 8.8 Impact of the number of seed bases on the area and time requirements

In order to verify the improvement in terms of performance, if any, $vs.$ the increase in terms of hardware area requirements occasioned by the use of more extension processors, we set $m = 20$, $n = 10$ and $w = 3$ and varied the number of processors p. Table 8.3 shows the impact as p increases. Fig. 8.9 illustrates graphically this impact.

Table 8.3 Hardware area and time requirements as p increases

p	FFs	%	LuTs	%	Slices	%	Time
1	6435	14	6622	15	3109	28	17.63
2	7887	18	7811	17	3300	29	12.59
3	7887	18	7804	17	3438	31	10.08
4	8004	18	7989	18	3672	33	8.50

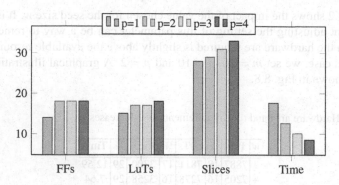

Fig. 8.9 Impact of the number of processor on the area and time requirements

Table 8.4 shows the time requirements of the MicroBlaze software implementation and the HBLAST hardware implementation. The operation frequency of processor MicroBlaze is 50 MHz while HBLAST runs at different frequencies as shown the penultimate column of Table 8.4. Fig. 8.10 illustrates, in a logarithmic scale, the comparison of the MicroBlaze and HBLAST performances, as well as the speedup achieved by using HBLAST. The average speedup is about $60\times$.

Table 8.4 Microblaze *vs* HBLAST time comparison

Case	m	n	Microblaze		HBLAST		
			#Cycles	Time	#Cycles	Freq.	Times
1	20	10	32411	528.59	772	61.3	12.59
2	60	20	54393	996.21	814	54.6	14.91
3	100	10	54919	1065.04	1009	51.5	19.58
4	100	25	255454	5109.08	3206	50.0	64.12

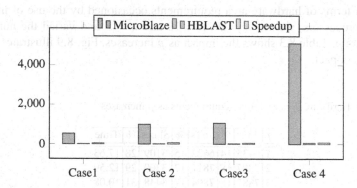

Fig. 8.10 Impact of the number of processor on the area and time requirements

8.5 Summary

This chapter presents a parallel architecture of the BLAST algorithm implemented as a hardware co-processor to the MicroBlaze processor. BLAST is used to align DNA sequences. The FPGA used is a Xilinx Virtex 5 FPGA (xc5vfx70t). The proposed architecture exploits the parallelism of identifying the seeds using a strategic partitioning of the subject and query sequences into words of a configurable size in terms of bases. It also explores further parallelism as it includes many extension processors to investigates the seeds found. The seeding and extension processes are carried on in a pipelined fashion.

Moreover, the design is easily scalable to new configuration parameter, which consist of the seed size w in terms of number of bases and the number of extension processors p. This adjustment cab be done according to speed *vs.* cost constraints.

A thorough analysis of the impact of each of the algorithm parameters has been done to evaluate the impact in terms of hardware are and time requirements. A comparison of the software-based and the proposed hardware design showed that a speedup of $60\times$ is achieved in average.

Future work will be directed at completing the assessment step and analyzing the impact on the whole design, as well as the use of real-world cases DNA alignment and matching.

References

1. Altschul, S., Gish, W., Miller, W., Myers, E.W., Lipman, D.J.: Basic local alignment search tool. Journal of Molecular Biology 215(3), 403–413 (1990)
2. Mount, D.W.: Steps used by the BLAST algorithm. Cold Spring Harbor Protocols: Molecular Biology (2007), doi:10.1101/pdb.ip41
3. Needlman, S., Wunsh, S.: A general method applicable to the search of similarities in Amino-Acid sequence of two protein. Journal of Molecular Biology 1(48), 443–453 (1970)
4. Garcia Neto Segundo, E.J., Nedjah, N., de Macedo Mourelle, L.: A parallel architecture for DNA matching. In: Xiang, Y., Cuzzocrea, A., Hobbs, M., Zhou, W. (eds.) ICA3PP 2011, Part II. LNCS, vol. 7017, pp. 399–407. Springer, Heidelberg (2011)
5. Giegerich, R.A.: Systematic approach to dynamic programming in bioinformatics. Bioinformatics 8(16), 665–677 (2000)
6. Kasap, S., Benkrid, K.: High performance phylogenetic analysis with maximum parsimony on reconfigurable hardware. IEEE Transactions on Very Large Scale Integration VLSI Systems 99(5), 796–808 (2011)
7. Pearson, W.: Searching protein sequence libraries: comparison of the sensitivity and selectivity of the Smith-Waterman and FASTA algorithms. Genomics 3(11), 635–650 (1991)
8. Rubin, E., Pietrokovski, S.: Heuristic methods for sequence alignment. Advanced Topics in Bioinformatics, Weizmann Institute of Science (2003)
9. Shaper, E.G., et al.: Sensitivity and selectivity in protein similarity searches: a comparison of Smith-Waterman in hardware to BLAST and FASTA. Genomics 2(38), 179–191 (1996)
10. Waterman, M.S.: Introduction to Computational Biology. CRC Press (1995)

Part II
Soft Computing for Hardware

Chapter 9
Synchronous Finite State Machines Design with Quantum-Inspired Evolutionary Computation*

Abstract. Synchronous finite state machines are very important for digital sequential designs. Among other important aspects, they represent a powerful way for synchronizing hardware components so that these components may cooperate adequately in the fulfillment of the main objective of the hardware design. In this chapter, we propose an evolutionary methodology based on the principles of quantum computing to synthesize finite state machines. First, we optimally solve the state assignment NP-complete problem, which is inherent to designing any synchronous finite state machines. This is motivated by the fact that with an optimal state assignment, one can physically implement the state machine in question using a minimal hardware area and response time. Second, with the optimal state assignment provided, we propose to use the same evolutionary methodology to yield an optimal evolutionary hardware that implements the state machine control component. The evolved hardware requires a minimal hardware area and imposes a minimal propagation delay on the machine output signals.

9.1 Introduction

Sequential digital systems or simply finite state machines (FSMs) have two main characteristics: there is at least one feedback path from the system output signal to the system input signals; and there is a memory capability that allows the system to determine current and future output signal values based on the previous input and output signal values [1].

Traditionally, the design process of a state machine passes through five main steps, wherein the second and third steps may repeated several times as shown in Figure 9.1.

1. the specification of the sequential system, which should determine the next states and outputs of every present state of the machine. This is done using state tables and state diagrams;

---

* This chapter was developed together with Marcos Paulo Mello Araujo.

N. Nedjah and L. de Macedo Mourelle, *Hardware for Soft Computing and Soft Computing for Hardware*, Studies in Computational Intelligence 529,
DOI: 10.1007/978-3-319-03110-1_9, © Springer International Publishing Switzerland 2014

2. the state reduction, which should reduce the number of present states using equivalence and output class grouping;
3. the state assignment, which should assign a distinct combination to every present state. This may be done using Armstrong-Humphrey heuristics [1, 2, 3];
4. the minimization of the control combinational logic using K-maps and transition maps;
5. finally, the implementation of the state machine, using gates and flip-flops.

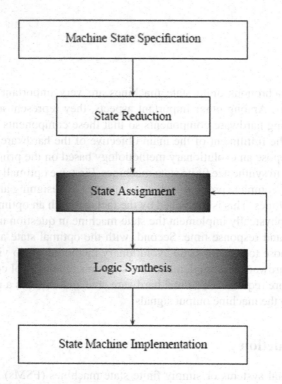

Fig. 9.1 Design methodology for sequential circuits

In this chapter, we concentrate on the third and forth steps of the design process, i.e. the state assignment and the control logic minimization problems. We present a quantum-inspired genetic algorithm designed to find a state assignment of a given synchronous finite state machine, which attempts to minimize the cost related to the state transitions. Then, we adapt the same quantum-inspired evolutionary algorithm to evolve the circuit that controls the machine current and next states.

The problems involved in state machine synthesis have been extensively studied in the past [2, 3, 1, 4]. These studies can be applied to state machine with a limited complexity, i.e. few state and transitions to control. Furthermore, the evolutionary principle in the form of genetic algorithms and genetic programming has been explored to solve these problems [5, 6, 7, 8]. The application of this principle allowed designers to synthesize more complex, state machines, i.e. with a little more states and transitions, and without much design effort. However, when the complexity of the state machine at hand goes beyond a certain limit, this applications fails to yield interesting synthesis results and also, the execution extends over hours of evolution One of the attractive properties of quantum computing is the possibility of massive parallelism, as it will be detailed later in the next sections of this chapter. This parallelism is not explicit. Instead, it is embedded within the information representation.

The use of both the evolutionary principle combined with that of quantum computing should allow us to improve further the synthesis process both in terms of improving the quality of the yielded results and also in synthesizing more complex state machine with no design effort and with shorter evolution time. The results presented towards the end of this chapter prove that the use of the quantum-inspired evolutionary process is very efficient. Using the proposed algorithm, we were able to synthesize automatically and evolutionary very complex state machines in a record time. In practical terms, our algorithm can be embedded in hardware synthesis tools to improve the quality of the synthesis result and generate those result efficiently.

The remainder of this chapter is organized into six sections. In Section 9.2, we introduce the problems that face the designer of finite state machine, which are mainly the state assignment problem and the design of the required control logic. In Section 9.3, we show that a well chosen assignment improves considerably the cost of the control logic. In Section 9.4, we give a thorough overview on the principles of quantum computing. In Section 9.5, we design a quantum-inspired genetic algorithm, which we call *QIGA* for evolving innovative solutions to hard *NP*-complete problems. In Section 9.6, we apply QIGA to the state assignment problem and we describe the genetic operators used as well as the fitness function, which determines whether a state assignment is better that another and how much. Subsequently, in Section 9.7, we present a quantum-inspired synthesizer for evolving efficient control logic circuit given the state assignment for the specification of the state machine in question. Then, we describe the circuit encoding, quantum gates used as well as the fitness function, which determines whether a control logic design is better than another and how much. Towards the end of this chapter, in Section 9.8, we present the results evolved by QIGA for some well-known FSM benchmarks. Then we compare the obtained results with those obtained by another genetic algorithm described in [7, 6] as well as with NOVA^{TM}, which uses well established but non-evolutionary methods [9]. We also provide the area and time requirements of the designs evolved through our evolutionary synthesizer for those benchmarks and compare the yielded results with those obtained using the traditional method to design state machines [9]. Last but no least, in Section 9.9, we draw some conclusions about this study and give some directions for future work.

9.2 Design Methodology of Synchronous Finite State Machines

Digital systems can be classified as *combinational systems* or *sequential systems*. A combinational system must obey the following restrictions [1]:

1. The values 0/1 of the output signals must depend on the actual values 0/1 of the input signals only.
2. There should be no feedback of the output signals to the input signals.

The aforementioned two restrictions make the design and analysis of combinational systems a straightforward task. Each output signal can be expressed as a Boolean function of the input signals. For a combinational system of n input signals and m output signals, we can have:

$$o_j = \phi_j(i_1, i_2, \ldots, i_n), \quad j = 1, 2, \ldots, m \tag{9.1}$$

wherein $i_1, i_2, \ldots, i_n$ area the input signals, $o_1, o_2, \ldots, o_m$ are the output signals and $\phi_1, \phi_2, \ldots, \phi_m$ form the necessary m Boolean function that yield the output signals.

In many digital systems, the output signal behavior cannot be determined knowing only the actual behavior of the input signals. In this case, the history of the input and output signals must be used to do so. Sequential systems are fundamentally different from the combinational ones, in spite of the fact that the former also include a combinational part. The term *sequential* is commonly used to describe this distinction. Sequential systems present two main characteristics:

1. There exists at least on path of feedback between the output and input signals of the system;
2. The circuit has the ability of remember information about the system past, in such a way that previous values of the output signals could be used to determine their respective next values.

The removal of the combinational restrictions allows for a larger spectrum for the application of digital systems. The use of memory elements and the feedback feature allow for the consideration of the time element as a parameter in the definition of the system behavior. Therefore, the information related to past events can be used to determine the behavior of the output signals. Moreover, information about both the past and the present can be captured as to plan and specify some future activities.

A clear advantage that can be observed through the comparison of sequential and purely combinational systems is the reduction of the hardware required due to the repetitive nature of sequential systems. However, a sequential system almost always requires more time to execute tasks [1]. The generic architecture of a Mealy finite state machine is given in Figure 9.2.

The input signals of a sequential system can be divided into two groups: *primary input signals* $(i_1, i_2, \ldots, i_n)$ and *secondary input signals* $(p_1, p_2, \ldots, p_k)$. The behavior of the primary input signals define the actual value of the system input, which can be one of the 2^n different possible combinations. The behavior of the secondary input signals reflects the past history of the sequential system. These signals are also called *current state signals* and whose values are read from the system memory.

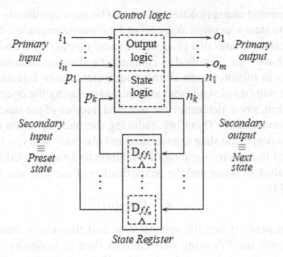

Fig. 9.2 A structural description of a Mealy state machine

The system ability to remember information about the past can be implemented through the utilization of *flip-flops* or *latches* [10]. The set of flip-flops used is generally called the *state register*. The k signal values of the secondary input form what is commonly known as the *present state* of the system. Therefore, the system may have 2^k distinct possible states. For this reason, sequential systems are also commonly called as *finite state systems* [10]. The *total state* of the system is defined as the union of the two sets of primary and secondary input signals. So, there are 2^{n+k} different total states.

The output signals can also be divided into two groups: *primary output signals* ($o_1, o_2, \ldots, o_m$) and *secondary output signals* ($n_1, n_2, \ldots, n_k$). The primary output signals form the control signals that are sent to the environment in which the sequential system is embedded. The secondary output signals form the data for the sequential system memory. These signals present the new value that will be saved into the system memory as soon as the next cycle of operation starts. Therefore, the secondary output signals are commonly called the *next state* of the system. In the same moment that the next state signals are written into the state register, the system passes to show this state as the present state. The primary and secondary output signals of the system are yield by combinational operations on the total state signals.

The design methodology of a state machine that controls the behavior of a given digital system may be subdivided into the following main steps:

• **Machine Specification:** The relationship between the present state signals and the primary input signals and that between the next state signals and the primary output signals describes the behavior of the sequential system. This relationship can be represented in many different ways. The most commonly used representations are the *state transition diagram* and the *state transition table*.

- **State Reduction:** States that produce the same output signal and have the same next state behavior are identified as *equivalent* and so are combined into a single state that acts in substitution to all these equivalent states. Equation 9.2 suggests that the total number of states that are necessary during the operation of the sequential system, say n, determine the minimal number of the state signals in that system implementation. Therefore, reducing the number of the included states yields a reduction in the state register size and also may lead to a reduction in the complexity of the control logic required. Some techniques used for the identification of equivalent states and the simplification of the state machine model can be found in [4].

$$K = \lceil \log_2(n) \rceil \qquad (9.2)$$

- **State Assignment:** Once the specification and the state reduction steps have been completed, the following step consists then of assigning a code to each state present in the machine. It is clear that if the machine has N distinct states then one needs N distinct combinations of 0s and 1s. So, one needs K flip-flops to store the machine current state, wherein K is the smallest positive integer such that $2^K \geq N$. The state assignment problem consists of finding the best assignment of the flip-flop combinations to the machine states. Since a machine state is nothing but a counting device, a combinational control logic is necessary to activate the flip-flops in the desired sequence. A generic architecture of a machine state is shown in Figure 9.2, wherein the feedback signals constitute the machine state, the control logic is a combinational circuit that computes the state machine primary output signals from the current state signals and the primary input signals. It also produces the signals of the machine next state.

  Let n be the number of states in a given machine and so $b = \lceil \log_2 n \rceil$ flip-flops are needed to store the machine state. A state assignment consists of identifying the 2^b binary codes that should be used to identify the machine n states. The number of possible distinct state assignments $f(n,b)$ [11] is given in Equation 9.3.

$$f(n,b) = \frac{2^b}{(2^b - n)} \qquad (9.3)$$

  Table 9.2 shows the values obtained for f when applied to some specific values of n and b. For instance, if the evaluation of an assignment as to its impact on the state machine implementation lasts say 100 μs, then 66 years would be needed to test all possible assignments, which cannot be done. Therefore, it is essential to use heuristics to overcome this problem.

- **Logic Synthesis:** The control logic component in a state machine is responsible for generating the primary output signals as well as the signal that form the next state. It does so using the primary input signals and the signals that constitute the current state (see Figure 9.2). Traditionally, the combinational circuit of the control logic is obtained using the transition maps of the flip-flops [1]. Given a state transition function, it is expected that the complexity, in terms of area and time,

Table 9.1 Number of possible state assignments

n	b	$f(n,b)$
2	1	2
3	2	24
4	2	24
5	3	6720
6	3	20160
7	3	40320
8	3	40320
9	4	$\approx 4 \cdot 10^9$
10	4	$\approx 3 \cdot 10^{10}$
11	4	$\approx 2 \cdot 10^{11}$
12	4	$\approx 9 \cdot 10^{11}$
13	4	$\approx 3 \cdot 10^{12}$
14	4	$\approx 1 \cdot 10^{13}$
15	4	$\approx 2 \cdot 10^{13}$

and so the cost of the control logic will vary for different assignments of flip-flop combinations to the allowed states. Consequently, the designer should seek the assignment that minimizes the complexity and so the cost of the combinational logic required to control the state transitions.

9.3 Impact of State Assignment

Given a state transition function, the requirements of area and time vary with respect to the state assignment used. Therefore, the designer or the computer-aided design tool for circuit synthesis needs always to select carefully the state assignment to be used. Existing techniques for state assignment can be listed as follows:

- *One-hot*: This technique associates a bit in the state register to each one of the existing state. This simplifies a great deal the synthesis flux as the control logic circuit can be obtained on-the-fly. However, it requires a register state whose size is defined by the number of states in the machine [10].
- *Heuristics*: These techniques attempt to identify a "good" assignment based on some heuristics. For instance, in [2] and [3], a heuristic based on state code adjacency, which attempts to assign adjacent codes to states that are "close" considering the state transition function. Two states are said to be *close* if one is the next state to the other and two binary codes are said to be *adjacent* if these are distinct in one single position. The idea behind this heuristic is the fact that adjacent binary codes will appear next to each other in Karnaugh maps and therefore would allow larger grouping, when necessary.

- *Meta-heuristics*: Evolutionary algorithms are used to evolve efficient assignments, rendering the assignment problem to an optimization one [6, 12]. These algorithms have been proven very efficient, very robust and the results obtained are far superior to those yield by the heuristic-based techniques

In order to demonstrate the impact of the chosen state assignment on the control logic complexity in terms of area and response time, let us consider the state machine described in Table 9.2 and try two different state codifications, which are $assignment_1 = \{00, 11, 01, 10\}$ and $assignment_2 = \{00, 01, 11, 10\}$. The circuit schematics for the state machine using $assignment_1$ and $assignment_2$ are shown in Figure 9.3 and 9.4 respectively.

Table 9.2 Example of state transition table

present state	next state		output (O)	
	$I = 0$	$I = 1$	$I = 0$	$I = 1$
s_0	s_0	s_1	0	0
s_1	s_2	s_1	0	1
s_2	s_0	s_3	1	0
s_3	s_2	s_1	1	1

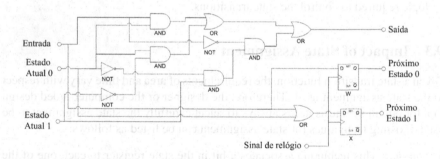

Fig. 9.3 Circuit schematics for the state machine using $assignment_1$

This example proves that the appropriate state assignment can reduce the implementation cost of the machine. The cost is defined here as the number of gates NOT, AND and OR of two one-bit inputs used. The inverted output signal of the flip-flops are considered of cost zero for the circuit implementation as these are available as output from the flip-flops. Assuming that the implementation cost of a given circuit is defined as the number of logic gates included, then Table 9.3 summarizes this cost for several possible state assignments, including $assignment_1$ and $assignment_2$. The afore-described example is an illustration of the fact that the choice of state assignment can reduce considerably the cost of state machine implementations, if chosen carefully.

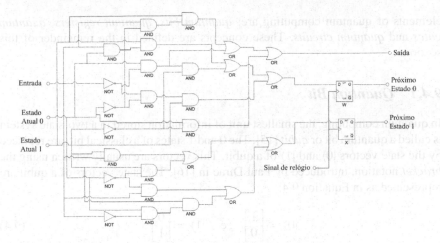

Fig. 9.4 Circuit schematics for the state machine using *assignment*₁

Table 9.3 Comparison of the number of logic gates for several possible state assignments

assignment	#AND	#OR	#NOT	Total
[00, 11, 01, 10]	4	3	1	8
[00, 01, 10, 11]	5	2	1	8
[00, 10, 01, 11]	5	2	1	8
[00, 11, 10, 01]	5	3	1	9
[11, 00, 01, 10]	5	3	1	9
[00, 01, 11, 10]	10	7	1	18
[00, 10, 11, 01]	11	6	1	18

In Section 9.6, we concentrate on the third step of the design process, i.e. the state assignment problem. We present a quantum-inspired genetic algorithm, designed for finding a state assignment of a given synchronous finite state machine, which attempts to minimize the cost related to the state transitions. In Section 9.7, we focus on evolving minimal control logics for state machines for a given state assignment and using an adapted version of the quantum-inspired genetic algorithm. Before getting to that, however, we first give an introduction to quantum computing and then we sketch the proposed algorithm.

9.4 Principles of Quantum Computation

Quantum computing is based on the concepts of quantum mechanics and is expected to be one of the main pillars of next generation computers. Many researchers are already using the principles of quantum computing to develop new techniques and algorithms to take advantage of the underlaying benefits [13, 14]. The basic elements of quantum computing are: *quantum bits, quantum registers, quantum gates* and *quantum circuits*. These concepts are defined in the remainder of this section.

9.4.1 Quantum Bit

In quantum computing, the smallest unit of information stored in a two-state system is called a quantum bit or *qubit* [15]. The 0 and 1 states of a classical bit, are replaced by the state vectors $|0\rangle$ and $|1\rangle$ of a qubit. This vectors are usually written using the *bracket* notation, introduced by Paul Dirac in [16]. The state vectors of a qubit are represented as in Equation 9.4:

$$|0\rangle = \begin{bmatrix} 1 \\ 0 \end{bmatrix} \quad e \quad |1\rangle = \begin{bmatrix} 0 \\ 1 \end{bmatrix}. \tag{9.4}$$

While the classical bit can be in only one of the two basic states that are mutually exclusive, the generic state of one qubit can be represented by a linear combination of the state vectors $|0\rangle$ and $|1\rangle$, as in Equation 9.5:

$$|\psi\rangle = \alpha |0\rangle + \beta |1\rangle, \tag{9.5}$$

wherein α and β are complex numbers. The state vectors $|0\rangle$ and $|1\rangle$ form a canonical base and the vector $|\psi\rangle$ represents the superposition of this vectors, with α and β amplitudes. The unit normalization of the state of the qubit ensures that Equation 9.6 is true:

$$|\alpha|^2 + |\beta|^2 = 1. \tag{9.6}$$

The phase of a qubit is defined by an angle ζ, defined as in Equation 9.7:

$$\zeta = \arctan(\beta/\alpha), \tag{9.7}$$

and tthe quadrant of qubit phase ζ is defined as in (9.8). If d is positive, the phase ζ lies in the first or third quadrant; otherwise, the phase ζ lies in the second or fourth quadrant [17].

$$d = \alpha \cdot \beta, \tag{9.8}$$

The physical interpretation of the qubit is that it may be simultaneously in the states $|0\rangle$ and $|1\rangle$, which allows for an infinite amount of information to be stored in state $|\psi\rangle$. However, during the act of observing the state of a qubit, it collapses to a single state, i.e. either $|0\rangle$ or $|1\rangle$ [18]. The qubit collapses to state $|0\rangle$, with probability $|\alpha|^2$ or state $|1\rangle$, with probability $|\beta|^2$.

9.4.2 Quantum Registers

A system with m qubits contains information on 2^m states. The linear superposition of possible states can be represented as in Equation 9.9:

$$|\psi\rangle = \sum_{k=1}^{2^m} C_k |S_k\rangle , \tag{9.9}$$

wherein C_k specifies the probability amplitude of the corresponding states $|S_k\rangle$ and subjects to the normalization condition of Equation 9.10.

$$|C_1|^2 + |C_2|^2 + ... + |C_{2^m}|^2 = 1 \tag{9.10}$$

9.4.3 Quantum Gates

The state of a qubit can be changed by the operation of a quantum gate or q-gate. The q-gates applies a unitary operation U on a qubit in the state $|\psi\rangle$, making it evolve to the state $U|\psi\rangle$, maintaining the probabilities interpretation defined in Equation 9.6. There are several q-gates, such as the *NOT* gate, *Controlled-NOT* gate, *Hadamard* gate, *rotation* gate [15].

9.5 Quantum-Inspired Genetic Algorithms

Since the emerging of evolutionary computation field, many new hybridized algo-rithms and technique based on the main concepts of evolution have been devel-oped. Just to name few, we can cite multi-objective evolutionary algorithms [20, 21], swarm-based techniques [19], differential evolution [22] and quantum-inspired evo-lutionary algorithm [23]. As any evolutionary algorithms, the latter is based on a population of solutions which is maintained through many generations. It seeks the best fitted solution to the problem, by evaluating the characteristics of those included in the current population. In the next section, we describe the quantum-inspired rep-resentation of the individual and the underlaying computational process.

9.5.1 Solution Representation

Evolutionary algorithms, like genetic algorithms, for instance, can use several rep-resentations that have been used with success: binary, integer, real or even symbolic [24]. The quantum-inspired evolutionary algorithms use a new probabilistic repre-sentation, that is based on the concept of qubits as defined in Equation 9.5 and q-individuals, which consist of a string of qubits. A q-individual, say p, can be viewed as in Equation 9.11, wherein $|\alpha_i|^2 + |\beta_i|^2 = 1$, for $i = 1, 2, 3, ..., m$.

$$p = \begin{bmatrix} \alpha_1 & \alpha_2 & \alpha_3 & \cdots & \alpha_m \\ \beta_1 & \beta_2 & \beta_3 & \cdots & \beta_m \end{bmatrix} \tag{9.11}$$

The advantage of the representation of the individuals using qubits instead of the classical representation of bits is the ability of representing the linear superpositions of all possible states. For instance, an individual represented with three qubits ($m = 3$) can be depicted as in Quation 9.12:

$$p = \begin{bmatrix} \frac{1}{\sqrt{2}} & \frac{1}{\sqrt{3}} & \frac{1}{2} \\ \frac{1}{\sqrt{2}} & \sqrt{\frac{2}{3}} & \frac{\sqrt{3}}{2} \end{bmatrix}, \quad (9.12)$$

or viewed in the alternative way of Equation 9.13,

$$p = \frac{1}{2\sqrt{6}}|000\rangle + \frac{1}{2\sqrt{2}}|001\rangle + \frac{1}{2\sqrt{3}}|010\rangle + \frac{1}{2}|011\rangle + \frac{1}{2\sqrt{6}}|100\rangle + \frac{1}{2\sqrt{2}}|101\rangle + \frac{1}{2\sqrt{3}}|110\rangle + \frac{1}{2}|111\rangle$$

$$(9.13)$$

The numbers in Equation 9.13 represent the amplitudes whose square-roots indicate the probabilities of observing states $|000\rangle$, $|001\rangle$, $|010\rangle$, $|011\rangle$, $|100\rangle$, $|101\rangle$, $|110\rangle$ and $|111\rangle$, which are $\frac{1}{24}, \frac{1}{8}, \frac{1}{24}, \frac{1}{12}, \frac{1}{24}, \frac{1}{8}, \frac{1}{24}$ and $\frac{1}{12}$, respectively.

The evolutionary algorithms with the quantum-inspired representation of individuals should permit a population diversity better than other representations, since the included individuals can represent linear superpositions of all possible states [25, 23]. For instance, the single q-individual of Equation 9.12 is enough to represent eight states. When using the classical representation of bits, eight individuals would be necessary to encode the same information.

9.5.2 Algorithm Description

The basic structure of the quantum-inspired evolutionary algorithm used in this chapter is described by Algorithm 9.1 [26].

The quantum-inspired evolutionary algorithm maintains a population of q-individuals, $P(g) = \{p_1^g, p_2^g, ..., p_n^g\}$ at generation g, where n is the size of population, and p_j^g is a q-individual defined as in Equation 9.14:

$$p_j^g = \begin{bmatrix} \alpha_{j_1}^g & \alpha_{j_2}^g & \alpha_{j_3}^g & \cdots & \alpha_{j_m}^g \\ \beta_{j_1}^g & \beta_{j_2}^g & \beta_{j_3}^g & \cdots & \beta_{j_m}^g \end{bmatrix}, \quad (9.14)$$

where m is the number of qubits, which defines the string length of the q-individual, and $j = 1, 2, ..., n$.

The initial population of n individuals is generated setting $\alpha_i^0 = \beta_i^0 = 1/\sqrt{2}$ ($i = 1, 2, ..., m$) of all $p_j^0 = p_j^g|_{g=0}$ for $j = 1, 2, ..., n$. This allows each q-individual to be the superposition of all possible states with the same probability.

The binary solutions in S_g are obtained by an observation process of the states of every q-individual in P_g. Let $S_g = \{s_1^g, s_2^g, ..., s_n^g\}$ at generation g. Each solution, s_i^g for $i = 1, 2, ..., n$, is a binary string with the length m, that is, $s_i^g = s_1 s_2...s_m$, where s_j is either 0 or 1.

Algorithm 9.1. Quantum-Inspired Genetic Algorithm – QIGA

$g := 0$;
generate P_0 with n individuals
observe P_0 into S_0
evaluate the fitness of every solution in S_0
store S_0 into B_0
while (**not** *termination condition*) **do**
    $g := g+1$;
    observe P_{g-1} into S_g
    evaluate the fitness of every solution in S_g
    update P_g using a q-gate and apply probability constraints
    store best solutions of B_{g-1}, S_g in B_g
    store the best solution in B_g into b
    if (*no improvement for many generation*) **then**
        replace all the solution of B_g by b
    end if
end while
return b

The observation process is implemented using random probability: for each pair of amplitudes $[\alpha_k, \beta_k]^T$ for $k = 1,2,...,n \times m$ of every qubit in the population P_g, a random number r in the range $[0,1]$ is generated. If $r < |\beta_k|^2$, the observed qubit is 1; otherwise, it is 0.

The q-individuals in P_g are updated using a q-gate, which is detailed later in the next section. We impose some probability constraints such that the variation operation performed by the q-gate avoid a premature convergence of a qubit to either to 0 or 1. This is done by allowing neither of $|\alpha|^2$ nor $|\beta|^2$ to reach 0 or 1. For this purpose, the probability $|\alpha|^2$ and $|\beta|^2$ are constrained to 0.02 as a minimum and 0.98 as a maximum. Such constraints allowed the algorithm to escape local minimum. This variation is one of the contribution of the chapter and has not been introduced in the original version of the algorithm [23].

After a given number of generations, if the best solution b does not improve, all the solutions stored into B_g are replaced by b. This step can induce a variation of the probabilities of the qubits within the q-individuals. This operation is also performed in order to escape local minimum and avoid the stagnant state.

9.6 State Assignment with QIGA

The identification of a good state assignment has been thoroughly studied over the years. In particular, Armstrong [2] and Humphrey [3] have pointed out that an assignment is good if it respects three rules, which consist of the following:

- two or more machine states that have the same next state should be given adjacent binary codes;

- two or more states that are the next states of the same state should be given adjacent binary codes.
- the first rule should have precedence over the second.

State adjacency means that the states appear next to each other in the mapped representation. In other terms, the combination assigned to the states should differ in only one position;

Now we concentrate on the assignment encoding and the fitness function. Given two different state assignments, the fitness function allows us to decide which is fitter.

9.6.1 State Assignment Encoding

In this case, a q-individual represents a state assignment. Each q-individual consists of an array of $2 \times N \lceil (\log_2 N) \rceil$ entries, wherein each set of $2 \times \lceil \log_2 N \rceil$ entries are the qubits associated to a single machine state. For instance, Figure 9.5 represents a q-individual and a possible assignment for a machine with 4 states obtained after the observation of the qubits.

S_0		S_1		S_2		S_3	
α_1^0	α_2^0	α_1^1	α_2^1	α_1^2	α_2^2	α_1^3	α_2^3
β_1^0	β_2^0	β_1^1	β_2^1	β_1^2	β_2^2	β_1^3	β_2^3

1	1	0	1	0	0	1	0

Fig. 9.5 Example of state assignment encoding

Note that when an observation occurs, one code might be used to represent two or more distinct states. Such a state assignment is not possible. In order to discourage the selection of such an assignment, we apply a penalty every time a code is used more than once within the considered assignment. This will be further discussed in the next section where the fitness function is described.

9.6.2 Q-Gate for State Assignment

To drive the individuals towards better solutions, a q-gate is used as a variation operator of the quantum-inspired evolutionary algorithm presented at this chapter. After an update operation, the qubit must always satisfy the normalization condition $|\alpha'|^2 + |\beta'|^2 = 1$, where α' and β' are the amplitudes of the updated qubit.

Initially, each q-individual represents all possible states with the same probability. As the probability of every qubit approaches either 1 or 0 as a result of many applications of the q-gate, the q-individual converges to a single state and the diversity property disappears gradually. By this mechanism, the quantum-inspired evolutionary algorithm can treat the balance between exploration and exploitation [23]. The q-gate used is inspired by a quantum rotation gate. This is defined in Equation 9.15.

$$
\begin{bmatrix} \alpha' \\ \beta' \end{bmatrix} = \begin{bmatrix} cos(\Delta\theta) & -sin(\Delta\theta) \\ sin(\Delta\theta) & cos(\Delta\theta) \end{bmatrix} \begin{bmatrix} \alpha \\ \beta \end{bmatrix},
\tag{9.15}
$$

where $\Delta\theta$ is the rotation angle of each qubit towards either of the states 0 or 1, depending on the amplitude signs. The angle $\Delta\theta$ should be adjusted according to problem at hand.

The value of the angle $\Delta\theta$ can be selected from the Table 9.4, where $f(s_i^g)$ and $f(b_i^g)$ are the fitness values of s_i^g and b_i^g, and s_j and b_j are the jth bits of the observed solutions s_i^g and the best solutions b_i^g, respectively. The rotation gate allows changing the amplitudes of the considered qubit, as follows:

1. If s_j and b_j are 0 and 1, respectively, and if $f(s_i^g) \geq f(b_i^g)$ is false then:

   - if the qubit is located in the first or third quadrant as defined in Equation 9.8, $\Delta\theta = \theta_3$ is set to a positive value to increase the probability of the state $|1\rangle$;
   - if the qubit is located in the second or fourth quadrant, $\Delta\theta = -\theta_3$ should be used to increase the probability of the state $|1\rangle$.

2. If s_j and b_j are 1 and 0, respectively, and if $f(s_i^g) \geq f(b_i^g)$ is false:

   - if the qubit is located in the first or third quadrant, $\Delta\theta = \theta_5$ is set to a negative value to increase the probability of the state $|0\rangle$;
   - if the qubit is located in the second or fourth quadrant, $\Delta\theta = -\theta_5$ should be used to increase the probability of the state $|0\rangle$.

Table 9.4 Look-up table of $\Delta\theta$

s_j	b_j	$f(s_i^g) \geq f(b_i^g)$	$\Delta\theta$
0	0	false	θ_1
0	0	true	θ_2
0	1	false	θ_3
0	1	true	θ_4
1	0	false	θ_5
1	0	true	θ_6
1	1	false	θ_7
1	1	true	θ_8

When it is ambiguous to select a positive or negative number for the angle parameter, we set its value to zero as recommended in [23]. The magnitude of $\Delta\theta$ has an effect on the speed of convergence. If it is too big, the search grid of the algorithm would be large and the solutions may diverge or converge prematurely to a local optimum. If it is too small, the search grid of the algorithm would be small and the algorithm may stagnate. Hence, the magnitude of $\Delta\theta$ varies and the corresponding values depend on the application problem. In the state assignment problem, we experimentally discovered that these values should be set as follows: $\theta_1 = \theta_2 = \theta_4 = \theta_6 = \theta_7 = \theta_8 = 0$, $\theta_3 = 0.05\pi$, and $\theta_5 = -0.05\pi$.

9.6.3 State Assignment Fitness

This step of the quantum-inspired evolutionary algorithm evaluates the fitness of each binary solutions obtained from the observation of the states of the q-individuals. The fitness evaluation of state assignments is performed with respect to the rules of Armstrong [2] and Humphrey [3]:

- how much a given state assignment adheres to the first rule, i.e. how many states in the assignment, which have the same next state but have no adjacent state codes;
- how much a given state in the assignment adheres to the second rule, i.e. how many states in the assignment, which are the next states of the same state but have no adjacent state codes.

In order to efficiently compute the fitness of a given state assignment, we use an $N \times N$ *adjacency matrix*, wherein N is the number of the machine states. The triangular bottom part of the matrix holds the expected adjacency of the states with respect to the first rule while the triangular top part of it holds the expected adjacency of the states with respect to the second rule. The matrix entries are calculated as described in Equation 9.16, wherein AM stands for the *Adjacency Matrix*, functions $next(\sigma)$ and $prev(\sigma)$ yield the set of states that are next and previous to state σ, respectively. For instance, the 4×4 adjacency matrix for the state machine presented in Table tab:estados is shown in Figure 9.6.

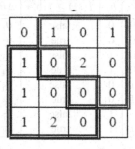

Fig. 9.6 Example of adjacency matrix

$$AM_{i,j} = \begin{cases} \#(next(q_i) \cup next(q_j)) & \text{If } i > j \\ \#(prev(q_i) \cup prev(q_j)) & \text{If } i < j \\ 0 & \text{If } i = j \end{cases} \qquad (9.16)$$

Using the adjacency matrix AM as defined in Equation 9.16, the fitness function applies a penalty of 2 or 1, every time the first or second rule are broken, respectively. The penalty of breaking first rule is higher than that associated with the second rule to maintain the higher priority of the former over the latter. Equation 9.17 shows the details of the fitness function applied to a state assignment σ, wherein function $na(q,p)$ returns 0 if the codes representing states q and p are adjacent and 1 otherwise. Note that if assignment σ associates two distinct states to the same binary code, the σ is penalized by adding the constant ψ to the corresponding fitness value.

$$f(\sigma) = \sum_{i \neq j \,\&\, \sigma_i = \sigma_j} \psi + \sum_{i=0}^{N-2} \sum_{j=i+1}^{N-1} (AM_{i,j} + 2 \times AM_{j,i}) \times na(\sigma_i, \sigma_j) \qquad (9.17)$$

For instance, considering the state machine whose adjacency matrix is described in Figure 9.6, the state assignment $\{s_0 \equiv 00, s_1 \equiv 10, s_2 \equiv 01, s_3 \equiv 11\}$ has a fitness of 5 as the codes of states s_0 and s_3 are not adjacent but $AM_{0,3} = AM_{3,0} = 1$ and the codes of states s_1 and s_2 are not adjacent but $AM_{1,2} = 2$ while the assignments $\{s_0 \equiv 00, s_1 \equiv 11, s_2 \equiv 01, s_3 \equiv 10\}$ has a fitness of 3 as the codes of states s_0 and s_1 are not adjacent but $AM_{0,1} = AM_{1,0} = 1$.

The objective of the quantum-inspired evolutionary algorithm is to find the assignment that minimizes the fitness function as described in Equation 9.17. Assignments with fitness 0 satisfy all the adjacency constraints. Note that such an assignment may not exist for some state machines.

9.7 Logic Synthesis with QIGA

Exploiting the quantum-inspired evolutionary algorithm, we can automatically generate novel control logic circuits that are reduced with respect to area and time requirements. The allowed gates are NOT, AND, OR, XOR, NAND, XNOR and WIRE, as shown in Table 9.5. The last row represents a physical wire and thus, the absence of a gate.

Table 9.5 Gate name, gate code, gate-equivalent and average propagation delay (ns)

Name	Code	Area	Delay
NOT	000	1	0.0625
AND	001	2	0.2090
OR	010	2	0.2160
XOR	011	3	0.2120
NAND	100	1	0.1300
NOR	101	1	0.1560
XNOR	110	3	0.2110
WIRE	111	0	0.0000

9.7.1 Circuit Codification

We encode circuit designs using a matrix of cells that may be interconnected. A cell
may or may not be involved in the circuit schematics and consists of two inputs, a
logical gate and a single output. A cell draws its input signals from the outputs of the
gates of the previous column. The cells located in the first column draw their inputs
from the circuit global input signals. Each cell is encoded with a number of qubits,
enough to represent the allowed gates and the signals that may be connected in each
input of the cell gate. Note that the total number of qubits may vary depending on
the number of outputs of the previous column and the number of primary inputs in
the case of the first column [27]. An example of a matrix of cells with respect to this
encoding is given in Figure 9.7.

For instance, the first part of Figure 9.8 represents a cell encoding and a possible
observation of the qubits states while the second part indicates the correspondent
circuit encoded by this cell, that is composed by an *AND* gate with its input A and
B connected to the first and third element of its previous column.

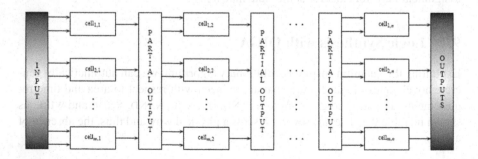

Fig. 9.7 Circuit representation

	Gate			Input A		Input B	
Cell	α_1 β_1	α_2 β_2	α_3 β_3	α_4 β_4	α_5 β_5	α_6 β_6	α_7 β_7
Observation	0	0	1	0	0	1	0

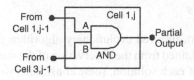

Fig. 9.8 Example of a cell considering that it has 4 outputs

When the observation of the qubits that define the gate yields 111, i.e. WIRE, then the signal connected to the cell's *A* input appears in the partial output of the cell. When the number of partial outputs of a column or the global inputs are not a power of 2, some of them are repeated in order to avoid that a cell be mapped to an inexistent input signal. The circuit primary output signals are the output signals of the cells in the last column of the matrix. If the number of global outputs are less than the number of cells in the last column, then some of the output signal are not used in the evolutionary process.

The power of the quantum-inspired representation can be evidenced in the drawing of Figure 9.9, which shows that all possible circuits can be represented with only one q-individual in a probabilistic way, as explained in the Section 9.5.1.

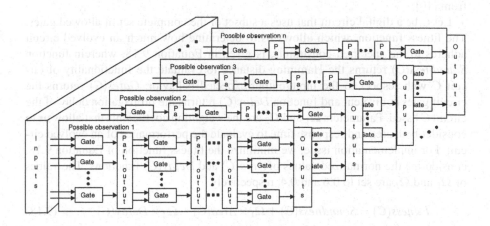

Fig. 9.9 Power of the quantum-inspired representation of an encoded circuit

The number of q-individual included in the population (population size) as well as the number of cells per q-individual are the parameters that should be adjusted considering the state machine complexity. The complexity depends on the number of inputs, outputs, states and number of states transitions of the machine.

9.7.2 Logic Fitness

This step of the quantum-inspired evolutionary algorithm evaluates the fitness of each binary solutions obtained from the observation of the states of the q-individuals. To evaluate the fitness of each solution, some constraints were considered: First of all, the evolved specification must obey the input/output behavior, which is given in a tabular form of the expected results given the inputs. This is the truth table of the expected circuit. Secondly, the circuit must have a reduced size. This constraint allows us to yield compact digital circuits. Finally, the circuit must also reduce the signal propagation delay. This allows us to reduce the response time and so discover efficient circuits.

We estimate the necessary area for a given circuit using the concept of gate-equivalent. This is the basic unit of measure for digital circuit complexity [10]. It is based upon the number of logic gates that should be interconnected to perform the same input/output behavior. This measure is more accurate that the simple number of gates [10].

When the input to an electronic gate changes, there is a finite time delay before the change in input is seen at the output terminal. This is called the propagation delay of the gate and it differs from one gate to another. We estimate the performance of a given circuit using the worst-case delay path from input to output. The number of gate-equivalent and an average propagation delay for each kind of gate were taken from [10].

Let C be a digital circuit that uses a subset or the complete set of allowed gates. The fitness function, which allows us to determine how much an evolved circuit adheres to the specified constraints, is given in Equation 9.18, wherein function $Soundness(C)$ returns the Hamming distance to evaluate the functionality of circuit C with respect to the input/output expected behavior, $Gates(C)$ returns the circuit gates equivalent and function $Delay(C)$ returns the propagation delay of the circuit C based. Parameters Ω_1 and Ω_2 are the weighting coefficients that allow us to consider both area and response time to evaluate the performance of an evolved circuit. For implementation issue, we minimize the fitness function of Equation 9.18, considering the normalized values of $Area(C)$ and $Delay(C)$ functions. The values of Ω_1 and Ω_2 are set to 0.6 and 0.4, respectively.

$$Fitness(C) = Soundness(C) + \Omega_1 \times Area(C) + \Omega_2 \times Delay(C), \qquad (9.18)$$

where the objective of QIGA is the minimization of this function.

The Hamming distance is an non-negative integer that is proportional to the number of errors that result from the comparison between the output of the evolved circuit and those expected for each of the possible combination of the input signals. Function *Soundness*(C) is in Equation 9.19. Note that this definition sums up a penalty ψ for each error and so the total value is proportional to the number of output signal that are different from the expected ones.

$$Soundness(C) = \sum_{i=1}^{p} |y_j - x_j| \times \psi \tag{9.19}$$

wherein p is the number of possible combinations of the input signals, $|y_j - x_j|$ is the difference between the output signals of the evolved circuit and the expected ones, i.e. x_j e y_j respectively and ψ is a constant penalty for a single error. Note that if *Soundness*(C) > 0 then the circuit does not implement the desired behavior correctly and therefore, this is considered as a penalty for the individuals that encode circuit C.

Function *Area*(C) returns the necessary hardware area to implement circuit C, which is evaluated using the number of gate-equivalent used. Let C be a circuit whose geometry is represented by a matrix $n \times m$. Recall that each cell $c_{i,j}$ of the circuit is formed by the gate type p together with the two inputs e_a e e_b. Function *Area*(C) is defined in Equation 9.20. This definition is expressed using a recursive function $Area_{i,j}$, which allows us to compute the required area by the portion of circuit C that produces the output of the gate at cell $c_{i,j}$. This function is defined in Equation 9.21. Note that the area corresponding to the shared gates must only be counted once. For this purpose, a Boolean matrix $V : n \times m$ whose entry $V_{i,j}$ is updated when the gate of cell $c_{i,j}$ has been visited. In Equation 9.21, $GE_{c_{i,j}^p}$ represents the number of gate-equivalent for gate p at cell $c_{i,j}$ and $c_{i,j}^{ex}$ represents one of the inputs of that gate.

$$Area(C) = \sum_{i=1}^{s} Area_{i,m}, \tag{9.20}$$

wherein s is the number of output signals of C with $s \leq m$.

$$Area_{i,j} = \begin{cases} GE_{c_{i,j}^p} & \text{If } j=1 \\ GE_{c_{i,j}^p} + \sum_{x \in \{a,b\} e \neg V_{c_{i,j}^{ex}, j-1}} \left(Area_{c_{i,j}^{ex}, j-1} \right) & \text{If } j \in [2,m] \end{cases} \tag{9.21}$$

When the input of a given gate switches from 0 to 1 or 1 to 0, there exists a finite delay before the change is perceived at the output terminal of that gate. This delay is called *propagation delay* and it depends on the type of the gate, the technology used to implement it and the load factor that is put on the output terminal of this gate. The values of the gate propagation delays for CMOS technology are given in Table 9.6, where L represents the total load on the gate output. This delay does also depend on the signal transition, i.e. the propagation delay of a gate are different when a positive (t_{pLH}) or negative (t_{pHL}) transition occurs. The total load for a given gate is based on a basic load unit defined for each gate family. The total load is then a sum of all

Table 9.6 Gates, respective delays, load factor and area

Gate Type	Propagation Delay		Load factor	Area
	t_{pLH}(ns)	t_{pHL}(ns)	(load unit)	(gate-equivalent)
NOT	$0.02 + 0.038L$	$0.05 + 0.017L$	1.0	1
AND	$0.15 + 0.037L$	$0.16 + 0.017L$	1.0	2
OR	$0.12 + 0.037L$	$0.20 + 0.019L$	1.0	2
XOR	$0.30 + 0.036L$	$0.30 + 0.021L$	1.1	3
NAND	$0.05 + 0.038L$	$0.08 + 0.027L$	1.0	1
NOR	$0.06 + 0.075L$	$0.07 + 0.016L$	1.0	1
XNOR	$0.30 + 0.036L$	$0.30 + 0.021L$	1.1	3

the load factor of every gate whose input signals is drawn from the output signal of the considered gate.

Let C be a circuit whose geometry is represented by a matrix $n \times m$. The delay introduced by cell $c_{i,j}$ is defined as in Equation 9.22, wherein $\alpha_{c_{i,j}^p}$ represents the average of the intrinsic delay of gate p at cell $c_{i,j}$. The average delay of the gate when the total load is 0 and $\beta_{c_{i,j}^p}$ the average delay due to the fanout of output signal of gate p of that cell. Table 9.7 shows the values of α and β for each of the used gates.

$$\tau gate_{i,j} = \alpha_{c_{i,j}^p} + \beta_{c_{i,j}^p} \times \left(\sum_{\substack{k \in [1,n], x \in \{a,b\}| \\ c_{k,j+1}^{ex} = i}} factor(c_{k,j+1}^p) \right) \tag{9.22}$$

Table 9.7 Values of α and β for the gates used by QIGA

Gate Type	α	β
NOT	0.035	0.0465
AND	0.155	0.0270
OR	0.160	0.0280
XOR	0.300	0.0285
NAND	0.065	0.0325
NOR	0.065	0.0455
XNOR	0.300	0.0285

The propagation delay of a circuit is defined by the delay of its critical path. Considering all possible paths in a circuit, the critical path is the one that yields the largest delay. The propagation delay of a given path of a circuit is defined by the sum of delay of each of the gates that is traversed by the signal from the input until the output of the circuit, as defined formally in Equation 9.23.

$$\tau path_{i,j} = \begin{cases} \tau gate_{i,j} & \text{If } j=1 \\ \tau gate_{i,j} + \max_{x \in \{a,b\}} \left(\tau path_{c_{i,j}^{ex}, j-1} \right) & \text{If } j \in [2,m] \end{cases} \tag{9.23}$$

For a circuit of $s \leq n$ output signals, the propagation delay is determined by the largest delay among those imposed by all the paths of the circuit that reach the s gates located at the last column of the matrix representing the circuit. Function $Delay(C)$ is then defined as in Equation 9.24.

$$Delay(C) = \max_{i \in [1,s]} \tau path_{i,m} \tag{9.24}$$

9.8 Performance Results

This section is divided into two main parts: the result evolved by QIGA for the state assignment problem and those obtained for the synthesis of the control logic. The FSMs used are well-known benchmarks for testing finite state machines [28].

9.8.1 State Assignments Results and Discussion

In this section, we compare the assignment evolved by the quantum-inspired evolutionary algorithm presented in this chapter to those yield by the genetic algorithms [12, 6] and to those obtained using the non-evolutionary assignment system called NOVA. Table 9.8 shows the best state assignments generated by the compared systems.

The graphs presented in Figure 9.10 – Figure 9.14 show the progress of the evolutionary process of the best assignment fitness together with the average fitness with respect to all individuals of the population for some of the state machines used in the comparison.

The results introduced in Table 9.8 are depicted in the charts of Figure 9.15 for the comparison of the gate number, Figure 9.16 for the comparison of the hardware area and Figure 9.17 for the comparison of the propagation delays.

In order to determine whether the results obtained by QIGA are significantly better than those obtained by the genetic algorithm and the NOVA™ synthesis tool, we performed a statistical test of significance. The most commonly used method of comparing proportions uses the χ^2-test [29]. This test makes it possible to determine whether the difference existing between two groups of data is significant or just a chance occurrence.

Table 9.8 Best state assignments found by the compared methods

FSM	Method	State Assignments
bbara	AG$_1$	[0,6,2,14,4,5,13,7,3,1]
	AG$_2$	[0,6,2,14,4,5,13,7,3,1]
	NOVA™	[9,0,2,13,3,8,15,5,4,1]
	QIGA	[4,5,1,9,13,12,14,15,7,6]
bbsse	AG$_2$	[0,4,10,5,12,13,11,14,15,8,9,2,6,7,3,1]
	NOVA™	[12,0,6,1,7,3,5,4,11,10,2,13,9,8,15,14]
	QIGA	[5,3,11,7,9,6,14,10,8,12,4,1,0,2,13,15]
dk14	AG$_1$	[5,7,1,3,6,0,4]
	AG$_2$	[0,4,2,1,5,7,3]
	NOVA™	[1,4,0,2,7,5,3]
	QIGA	[5,7,4,0,6,3,1]
dk16	AG$_1$	[12,8,1,27,13,28,14,29,0,16,26,9,2,4,3,10,11,17,24,5,18,7,21,25,6,20,19]
	NOVA™	[12,7,1,3,4,10,23,24,5,27,15,16,11,6,0,20,31,2,13,25,21,14,18,19,30,17,22]
	QIGA	[14,30,22,6,4,5,13,25,18,20,31,9,10,26,23,28,29,7,15,3,16,8,21,17,1,11,24]
donfile	AG$_1$	[0,12,9,1,6,7,2,14,11,,17,20,23,8,15,10,16,21,19,4,5,22,18,13,3]
	NOVA™	[12,14,13,5,23,7,15,31,10,8,29,25,28,6,3,2,4,0,30,21,9,17,12,1]
	QIGA	[7,6,23,31,26,27,15,14,13,5,10,4,22,30,12,8,11,9,18,19,2,0,3,1]
lion9	AG$_2$	[0,4,12,13,15,1,3,7,5]
	NOVA™	[2,0,4,6,7,5,3,1,11]
	QIGA	[11,9,3,1,2,0,8,10,14]
mod12	AG$_1$	[0,8,1,2,3,9,10,4,11,12,5,6]
	NOVA™	[0,15,1,14,2,13,3,12,4,11,5,10]
	QIGA	[15,7,6,14,10,2,3,1,5,13,9,11]
shiftreg	AG$_1$	[0,2,5,7,4,6,1,3]
	AG$_2$	[0,2,5,7,4,6,1,3]
	NOVA™	[0,4,2,6,3,7,1,5]
	QIGA	[4,0,2,6,5,1,3,7]
train11	AG$_2$	[0,8,2,9,13,12,4,7,5,3,1]
	NOVA™	[0,8,2,9,1,10,4,6,5,3,7]
	QIGA	[9,11,13,3,1,2,0,12,8,5,4]

For the sake of completeness, we explain briefly how the test works. χ^2-test determines the differences between the observed and expected measures. The observed values are the actual experimental results, whereas the expected ones refer to the hypothetical distribution based on the overall proportions between the two compared algorithms if these are alike. Let $\lambda_o^{(a,m,q)}$ and $\lambda_e^{(a,m,q)}$ be respectively the observed and expected value of objective q obtained when using algorithm a with machine state m. Note that $\lambda_e^{(a,m,q)}$ is computed as described in Equation 9.25.

$$\lambda_e^{(a,m,q)} = \frac{\sum_{(x,z)\in A\times Q} \lambda_o^{(x,m)} \times \sum_{y\in M} \lambda_o^{(a,y)}}{\sum_{(x,y,z)\in A\times M\times Q} \lambda_o^{(x,y,x)}}, \qquad (9.25)$$

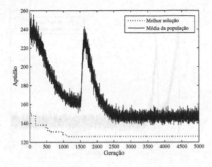
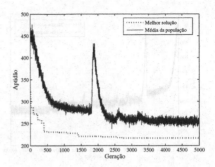

Fig. 9.10 Progress of the best solution fitness together with the average fitness for state machines *bbara* e *bbsse*

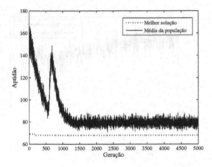
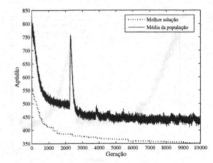

Fig. 9.11 Progress of the best solution fitness together with the average fitness for state machines *dk14* e *dk16*

wherein $A=\{$QIGA, NOVA$^{\text{TM}}\}$, $A=\{$QIGA, AG$_2\}$ or $A=\{$QIGA, AG$_3\}$, $M \subseteq \{$*bbara*, *bbsse*, *dk14*, *dk16*, *donfile*, *lion9*, *modulo12*, *shiftreg*, *train11*$\}$ and $Q=\{$*#gate*, *area*, *time*$\}$. The χ^2-test is based on the value of χ^2 computed as in Equation 9.26, wherein set $a = \{$QIGA $\times$ NOVA^{TM}, QIGA $\times$ AG$_1$ e QIGA $\times$ AG$_2\}$ and $q = \{$*#gate*, *area*, *time*$\}$.

$$\chi^2 = \sum_{(a,m,q)\in A \times M \times Q} \frac{\left(\lambda_o^{(a,m,q)} - \lambda_e^{(a,m,q)}\right)^2}{\lambda_e^{(a,m,q)}}. \tag{9.26}$$

The computed values for χ^2 for each of the comparisons are given in Table 9.9. The use of the χ^2-test is recommended when the proportions are small. Therefore, the time quantities were converted to 0.1 ns instead of 1 ns, thus avoiding the limitation imposed for the usage of the test.

The critical value of χ^2 is 0.05 (i.e. 95% of confidence) and considered the limit to assume the tested hypothesis. The degree of freedom depends on the amount

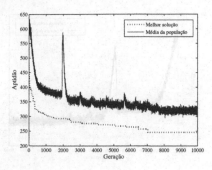 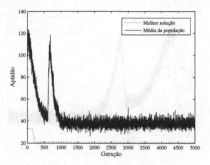

Fig. 9.12 Progress of the best solution fitness together with the average fitness for state machines *donfile* e *lion9*

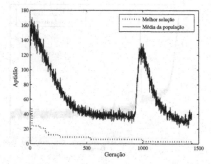 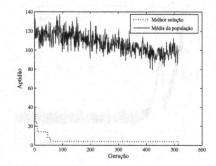

Fig. 9.13 Progress of the best solution fitness together with the average fitness for state machines *modulo12* e *shiftreg*

Table 9.9 Degree of freedom, computed χ^2, critical χ^2 for the confidence level of 99,5% e the degree of confidence obtained for the considered comparisons

Comparison	Degree of freedom	χ^2	Critical value	Confidence level
QIGA $\times$ NOVA^{TM}	40	73,302	66,766	>99,5%
QIGA $\times$ AG$_1$	25	68,281	46,928	>99,5%
QIGA $\times$ AG$_2$	30	64,740	53,672	>99,5%

of results used to compute χ^2. Assuming that the results are organized in a two-dimensional array of r rows and c columns, the degree of freedom is defined by $(r-1) \times (c-1)$. In this comparison, the number of rows coincides with that of state machines used as benchmarks and the number of columns is 6: one for each pair of objective/algorithm (we are considering 3 objectives and 2 algorithms in each comparison).

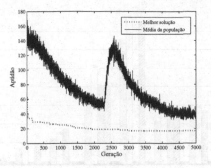

Fig. 9.14 Progress of the best solution fitness together with the average fitness for state machines *train11*

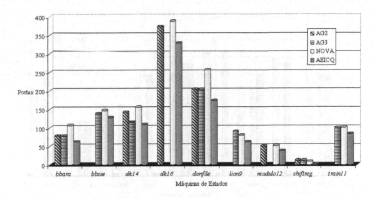

Fig. 9.15 Logic control comparison in terms of gates used

In the case of the comparison QIGA × NOVA™, all 9 state machines listed in Table 9.8 are considered, while in the case of the other two comparisons, i.e. QIGA × AG$_1$ and QIGA × AG$_2$, only some of the machines, 6 and 7 respectively, are used taking into account the results availability. At the light of the statistical analysis, we can conclude that QIGA performs significantly better than NOVA™, AG$_1$ e AG$_2$.

For all the simulations, we used a population of 50 q-individuals. However, we observed that for some state machines, such as *shiftreg* and *lion9*, the best solution was obtainable with a population of a single q-individual. Nevertheless, in this last case, the number of runs that reached the best result shrunk considerably. For instance, during the evolution of *shiftreg*, the global optimum was reached em all the runs when the population size was of 50 q-individuals while with a population of 1 q-individuals, this was the case for only in 50% of the runs. During the performed simulations, it was also possible to observe that QIGA was very robust with respect to the choice of the angle magnitudes θ_3 and θ_5 within the spectrum suggested in [23].

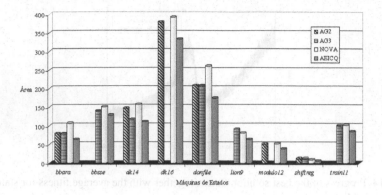

Fig. 9.16 Logic control comparison in terms of hardware area required

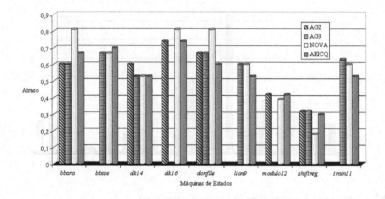

Fig. 9.17 Logic control comparison in terms of propagation delay imposed

The impact of the control phase of the probability amplitudes of the qubits, first contributed in QIGA, can be depicted in Figure 9.18. Figure 9.18–(a) shows that when the control is not imposed and the quantum-inspired algorithm does not evolve any new better solution, the average fitness of the population at hand gets very close to the fitness of the best q-individual, which has been yield so far. This happens due to the fact that the probabilities of the quantum states would practically be 100%, which would, in consequence lead to the measurement of the same solution in all generations. In contrast with this, Figure 9.18–(b) shows that the average of the probabilities is kept clear from the best solution. This is, actually, due to the control of the probability amplitudes of the qubits, which thus allows new solutions to be yield by the evolutionary process. This control step allows us to maintain a better diversity within the population individual and hopefully would accelerate the convergence of the optimization process.

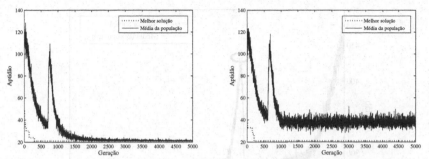

(a) without control of the probability ampli- (b) with control of the probability amplitudes
tudes

Fig. 9.18 Impact of the control phase of the probability amplitudes of the qubits

The impact of the global migration step can be viewed in Figure 9.19. In this step, the best solutions in $B(g)$, which are used in the update operation of the qubits, are all replaced by the best solution b. This substitution introduces a change in the population in the attempt to further improve its diversity. The picks, highlighted in the graphics of Figure 9.19, indicate three moments that allow for a clear observation of the effect caused by the global migration step on the average fitness of the population. The showed picks appear whenever 400 generations pass by without yielding a new better solution. Hence, as the control phase of the probability amplitudes of the qubits, this operation of global migration permits an remarkable improvement of the population diversity and thus leading to avoiding local minimums.

9.8.2 Logic Synthesis Results and Discussion

Table 9.10 shows the characteristics of the circuits that were synthesized using genetic programming (GP) [7, 8], genetic algorithms (GA) [6] and the ABC synthesis tool [9]. Table 9.11 shows the characteristics of the best circuit evolved by QIGA for each of the used machines.

The results listed in Table 9.11 and Table 9.10 are depicted as charts in Figure 9.20 for gate number comparison, Figure 9.21 for area comparison and Figure 9.22 for delay comparison.

The graphs presented in Figure 9.23 – Figure 9.27 show the progress of the evolutionary process of the best circuit fitness together with the average fitness with respect to all individuals in the population for some of the state machines used in the comparison.

As before, and in order to determine whether the results obtained by QIGA are significantly better than those obtained by GP [7, 8] and ABC [9]. The computed χ^2 for these comparisons are presented in Table 9.12.

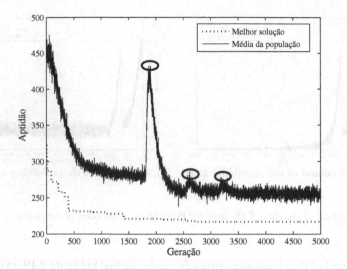

Fig. 9.19 Impact of the global migration

Table 9.10 Characteristics of evolved circuits by GP, GA, ABC

State machine	GP			GA			ABC		
	#Gates	Area	Delay	#Gates	Area	Delay	#Gates	Area	Delay
bbara	–	–	–	60	–	–	62	63	0,67
bbsse	–	–	–	–	–	–	128	128	0,70
bbtas	–	–	–	19	–	–	24	24	0,32
dk14	–	–	–	–	–	–	109	110	0,53
dk15	–	–	–	53	–	–	92	92	0,46
dk16									
dk27	–	–	–	16	–	–	25	25	0,32
dk512	–	–	–	47	–	–	63	63	0,46
donfile	–	–	–	–	–	–	174	174	0,60
lion9	21	39	0,70	50	–	–	62	63	0,53
modulo12	–	–	–	–	–	–	38	38	0,42
shiftreg	5	14	0,60	8	–	–	2	6	0,30
tav	–	–	–	26	–	–	31	31	0,46
train11	22	43	0,56	–	–	–	85	85	0,53

The property of *scalability* is of paramount importance for any kind of project and in electronic circuits projects, in particular [30, 31]. According to [32], scalability in evolutionary electronics can be approached in two different ways that are somehow related. The first focuses on the scalability of the individuals that represent electronic circuits. It was established that if no restriction on how basic components are connected is imposed then the size of the individuals will grow in the order of

Table 9.11 QIGA experimental results

State machine	#Gates	Area	Delay
bbara	**54**	78	0.88
bbtas	21	27	0.73
dk15	65	109	0.92
dk27	**15**	26	0.43
dk512	**47**	78	0.84
lion9	**20**	29	**0.52**
modulo12	19	34	0.56
shiftreg	**2**	2	**0.04**
tav	**26**	24	**0.32**
train11	25	37	**0.52**

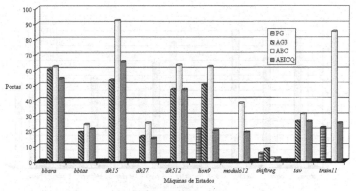

Fig. 9.20 Comparison of control logic for number of gates

Table 9.12 Degree of freedom, computed χ^2, critical χ^2 for the confidence level of 99,5% e the degree of confidence obtained for the considered comparisons

Comparison	Degree of freedom	χ^2	Critical value	Confidence level
QIGA × PG	10	18,898	18,31	>95,0%
QIGA × ABC	45	97,823	69,96	>99,5%

$O(n^2)$, wherein n is the number of functional components. However, if the connectivity is restricted to a local neighborhood in the proximity of the component, the order of $O(n)$ can be achieved. Note that the latter restricts the circuit that can be evolved. The second way to handle scalability in evolutionary circuits is to reduce, to a minimum, the complexity of the evolutionary process. Nowadays, scalability is the main problem that faces the extensive use of evolutionary electronics.

The problem of scalability was noted in many other works that used the evolutionary process to yield circuits [6, 33] and this was also the case for this work. For a sate machine of reduced complexity such as *shiftreg*, it is possible to encode the circuit with a 4 × 3 geometry composed of 108 qubits. In this case, the population

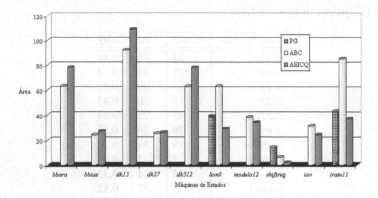

Fig. 9.21 Comparison of control logic for required area

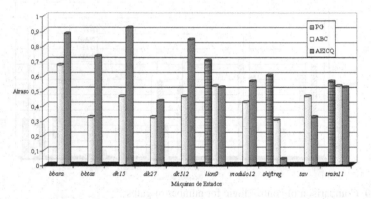

Fig. 9.22 Comparison of control logic for imposed delay

of 20 q-individuals is enough to yield optimal circuits. However, for more complex state machines, such as *bbara*, it is necessary to use a geometry of 32×5, composed of 2080 qubits. In this case, the search space becomes extremely large, dictating imperatively an increase of the population size, which in turn leads to a considerable increase of the average execution time. This time is about 3 minutes in the case of the *shiftreg* state machine and around 5 hours in the case of *bbara*.

The increase of required time of the evolution of circuits brings together an extra difficulty, which is the adjustment of the parameters needed in QIGA. This makes it inviable to refine the parameters considering the characteristics of the state machine at hand. To overcome this obstacle, we adjusted the parameter setup based on the state machines *lion9* and *train11* and these parameters were used during the evolution of the control logic of the remaining state machines. Even so, the results obtained for these machines are satisfactory, proving once again the robustness of QIGA.

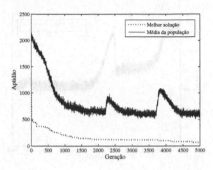
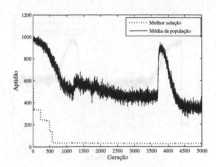

Fig. 9.23 Progress of the best solution fitness together with the average fitness for logic synthesis of state machines *bbara* e *bbtas*

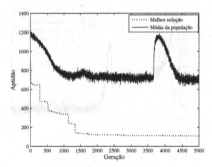
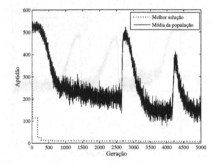

Fig. 9.24 Progress of the best solution fitness together with the average fitness for logic synthesis of state machines *dk15* e *dk27*

The results shown in this section suggest that QIGA is a tool of great potential to be used in automatic synthesis of electronic circuits. The evolved circuits show similar and some time better characteristics than those obtained by ABC, which is a well-known as a powerful tool for logic synthesis.

9.9 Summary

In this chapter we studied the application of quantum-inspired evolutionary methodology to solve two hard problems: the state assignment and the automatic synthesis of the control logic in the design process of synchronous finite state machines. We compared both the state assignment and the circuits evolved by the proposed algorithm QIGA for machines of different sizes and complexity with the results obtained by other method. QIGA almost always obtains better results. This proves that quantum-inspired evolutionary computation is very robust and leads to good results

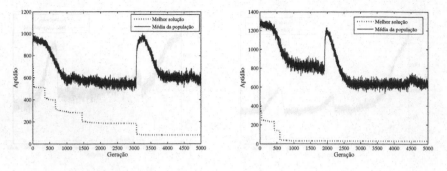

Fig. 9.25 Progress of the best solution fitness together with the average fitness for logic synthesis of state machines *dk512* e *lion9*

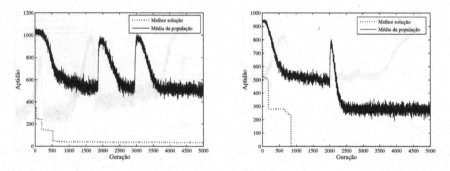

Fig. 9.26 Progress of the best solution fitness together with the average fitness for logic synthesis of state machines *modulo12* e *shiftreg*

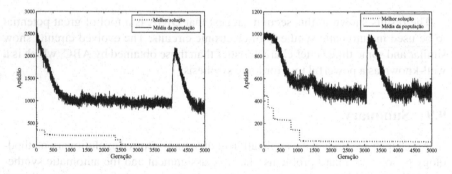

Fig. 9.27 Progress of the best solution fitness together with the average fitness for logic synthesis of state machines *tav* e *train11*

and therefore can be very profitable when embedded in automatic synthesis tools used in the design of digital systems.

Two main directions for future work emerges from this study. Regarding the state assignment problem, one can investigate the use of other heuristics other than or combined with Armstrong and Humphrey's [2, 3]. Regarding the logic synthesis, one can study the adaptation of QIGA so that it can evolve circuits at function level, instead of gate level as it is the case in this chapter. Another interesting investigation is the use of co-evolution technique in QIGA to accelerate further the evolutionary process. This allows one to catch up with the scalability problem.

References

1. Rhyne, V.T.: Fundamentals of digital systems design. In: Computer Applications in Electrical Engineering Series. Prentice-Hall (1973)
2. Armstrong, D.B.: A programmed algorithm for assigning internal codes to sequential machines. IRE Transactions on Electronic Computers EC-11(4), 466–472 (1962)
3. Humphrey, W.S.: Switching circuits with computer applications. McGraw-Hill, New York (1958)
4. Booth, T.L.: Sequential machines and automata theory. John Wiley & Sons, New York (1967)
5. Ali, B., Kalganova, T., Almaini, A.E.: Extrinsic evolution of finite state machine. In: Proc. of International Conference on Adaptive Computing in Design and Manufacture, pp. 157–168. Springer (2002)
6. Ali, B.: Evolutionary algorithms for synthesis and optimization of sequential logic circuits. Ph.D. Thesis, School of Engineering of Napier University, Edinburgh, UK (2003)
7. Nedjah, N., Mourelle, L.M.: Evolvable machines: theory and practice. STUD FUZZ, vol. 161. Springer, Heidelberg (2005)
8. Nedjah, N., Mourelle, L.M.: Mealy finite state machines: an evolutionary approach. International Journal of of Innovative Computing, Information and Control 2(4), 789–806 (2006)
9. ABC, A system for sequential synthesis and verification, Release 70930. In: Logic Synthesis and Verification Group, Berkeley (2005)
10. Ercegovac, M., Lang, T., Moreno, J.H.: Introduction to Digital Systems. John Wiley, USA (1998)
11. Hartmanis, J.: On the state assignment problem for sequential machines. IRE Transactions on Electronic Computers EC-10(2), 157–165 (1961)
12. Amaral, J.N., Tumer, K., Glosh, J.: Designing genetic algorithms for the state assignment problem. IEEE Transactions on Systems, Man, and Cybernetics 25(4), 686–694 (1995)
13. Shor, P.W.: Algorithms for quantum computation: discrete logarithms and factoring. In: Proc. the Annual Symposium on Foundations of Computer Science, pp. 124–134. IEEE Computer Society Press (1994)
14. Lin, F.T.: An enhancement of quantum key distribution protocol with noise problem. International Journal of of Innovative Computing, Information and Control 4(5), 1043–1054 (2008)
15. Hey, T.: Quantum computing. Computing Control Engineering Journal 10(3), 105–112 (1999)
16. Dirac, P.A.M.: The principles of quantum mechanics, 4th edn. Oxford University Press (1958)

17. Zhang, G.: Novel quantum genetic algorithm and its applications. Frontiers of Electrical and Electronic Engineering in China 1(1), 31–36 (2006)
18. Narayanan, A.: Quantum computing for beginners. In: Proc. of the Congress on Evolutionary Computation, vol. 3, pp. 2231–2238. IEEE Press, Piscataway (1999)
19. Nedjah, N., Mourelle, L.M. (eds.): Swarm Intelligent Systems. SCI, vol. 26. Springer, Heidelberg (2006)
20. Uno, T., Katagiri, H., Kato, K.: An evolutionary multi-agent based search method for stackelberg solutions of bi-level facility location problems. International Journal of of Innovative Computing, Information and Control 4(5), 1033–1042 (2008)
21. Liu, C., Wang, Y.: A new evolutionary algorithm for multi-objective optimization problems. ICIC Express Letters 1(1), 93–98 (2007)
22. Zhang, X., Lu, Q., Wen, S., Wu, M., Wang, X.: A modified differential evolution for constrained optimization. ICIC Express Letters 2(2), 181–186 (2008)
23. Han, K.H., Kim, J.H.: Quantum-inspired evolutionary algorithm for a class of combinatorial optimization. IEEE Transactions on Evolutionary Computation 6(6), 580–593 (2002)
24. Hinterding, R.: Representation, constraint satisfaction and the knapsack problem. In: Proc. of the Congress on Evolutionary Computation, vol. 2, pp. 1286–1292. IEEE Press, Piscataway (1999)
25. Akbarzadeh, M.R., Khorsand, A.R.: Quantum gate optimization in a meta-level genetic quantum algorithm. In: Proc. of IEEE International Conference on Systems, Man and Cybernetics, vol. 4, pp. 3055–3062. IEEE Press, Piscataway (2005)
26. Araujo, M.P.M., Nedjah, N., de Macedo Mourelle, L.: Optimised state assignment for fSMs using quantum inspired evolutionary algorithm. In: Hornby, G.S., Sekanina, L., Haddow, P.C. (eds.) ICES 2008. LNCS, vol. 5216, pp. 332–341. Springer, Heidelberg (2008)
27. Araujo, M.P.M., Nedjah, N., de Macedo Mourelle, L.: Logic synthesis for fSMs using quantum inspired evolution. In: Fyfe, C., Kim, D., Lee, S.-Y., Yin, H. (eds.) IDEAL 2008. LNCS, vol. 5326, pp. 32–39. Springer, Heidelberg (2008)
28. ACM/SIGDA, Collaborative Benchmarking and Experimental Algorithmic, North Carolina State University (2009), http://www.cbl.ncsu.edu
29. Diaconis, P., Efron, B.: Testing for independence in a two-way table: new interpretations of the chi-square statistic (with discussion). The Annals of Statistics 13, 845–913 (1985)
30. Higuchi, T.: Evolving hardware with genetic learning. In: Proc. of International Conference on Simulation Adaptive Behavior: A First Step Toward Building a Darwin Machine, pp. 417–424. MIT Press (1992)
31. Hemmi, H., Mizoguchi, J., Shimohara, K.: Development and evolution of hardware behaviors. In: Sanchez, E., Tomassini, M. (eds.) Towards Evolvable Hardware 1995. LNCS, vol. 1062, pp. 250–265. Springer, Heidelberg (1996)
32. Yao, X., Higuchi, T.: Promises and challenges of evolvable hardware. IEEE Transactions on Systems, Man, and Cybernetics, Part C: Applications and Reviews 29(1), 87–97 (1999)
33. Zebulum, R.S., Pacheco, M.A., Vellasco, M.M.: Evolutionary electronics: automatic design of electronic circuits and systems by genetic algorithms. CRC Press (2001)

Chapter 10
Application Mapping in Network-on-Chip Using Evolutionary Multi-objective Optimization*

Abstract. Network-on-chip (NoC) are considered the next generation of communication infrastructure, which will be omnipresent in most of industry, office and personal electronic systems. In the platform-based methodology, an application is implemented by a set of collaborating intellectual properties (IPs) blocks. In this chapter, we use multi-objective evolutionary optimization to address the problem of mapping topologically pre-selected sets IPs, which constitute the set of optimal solutions that were found for the IP assignment problem, on the tiles of a mesh-based NoC. The IP mapping optimization is driven by the area occupied, execution time and power consumption.

10.1 Introduction

As the integration rate of semiconductors increases, more complex cores for *system-on-chip* (SoC) are launched. A simple SoC is formed by homogeneous or heterogeneous independent components while a complex SoC is formed by interconnected heterogeneous components. The interconnection and communication of these components form a *network-on-chip* (NoC). A NoC is similar to a general network but with limited resources, area and power. Each component of a NoC is designed as an *intellectual property* (IP) block. An IP block can be of general or special purpose such as processors, memories and DSPs [4].

Normally, a NoC is designed to run a specific application. This application, usually, consists of a limited number of tasks that are implemented by a set of IP blocks. Different applications may have a similar, or even the same, set of tasks. An IP block can implement more than a single task of the application. For instance, a processor IP block can execute many tasks as a general processor does but a multiplier IP block for floating point numbers can only multiply floating point numbers. The number of IP blocks designers, as well as the number of available IP blocks, is growing up fast.

In order to yield an efficient NoC-based design for a given application, it is necessary to choose the adequate minimal set of IP blocks. With the increase of IP

---

* This chapter was developed in collaboration with Marcus Vinícius Carvalho da Silva.

N. Nedjah and L. de Macedo Mourelle, *Hardware for Soft Computing and Soft Computing for Hardware*, Studies in Computational Intelligence 529,
DOI: 10.1007/978-3-319-03110-1_10, © Springer International Publishing Switzerland 2014

blocks available, this task is becoming harder and harder. Besides IP blocks carefully assignment, it is also necessary to map the blocks onto the NoC available infra-structure, which consists of a set of *cores* communicating through *switches*. A bad mapping can degrade the NoC performance. Different optimization criteria can be pursued depending on how much information details is available about the application and IP blocks.

Usually, the application is viewed as a graph of tasks called *task graph* (TG). The IP blocks features can be obtained from their companion documentation. The IP assignment and IP mapping are key research problems for efficient NoC-based designs. These two problems are *NP*-hard problems and can be solved using multi-objective optimizations.

In this chapter, we propose a multi-objective evolutionary-based decision support system to help NoC designers. For this purpose, we propose a structured representation of the TG and an IP repository that will feed data into the system. We use the data available in the Embedded Systems Synthesis benchmarks Suite (E3S) [2] as our IP repository. The E3S is a collection of TGs, representing real applications based on embedded processors from the Embedded Microprocessor Benchmark Consortium (EEMBC). It was developed to be used in system-level allocation, assignment, and scheduling research. We used the NSGA-II, which is an efficient multiobjective algorithm that uses Pareto dominance as a selection criterion [1]. The algorithm was modified according to some prescribed NoC design constraints.

The rest of the chapter is organized as follows: First, in Section 10.2, we present briefly some related research work. Then, in Section 10.3, we introduce an overview of NoC structure. Subsequently, in Section 10.4, we describe a structured TG and IP repository model based on the E3S data. After that, in Section 10.5.1, we introduce the mapping problem in NoC-based environments. Then, in Section 10.5, we sketch the NSGA-II algorithm used in this work, individual representations and objective functions for the optimization stage. Later, in Section 10.7, we show some experimental result yield. Last but not least, in Section 10.8, we draw some conclusions and outline new directions for future work.

10.2   Related Work

The problems of mapping IP blocks into a NoC physical structure have been addressed in some previous studies. Some of these works did not take into account of the multi-objective nature of these problems and adopted a single objective optimization approach. Hu and Marculescu [4] proposed a branch and bound algorithm which automatically maps IPs/cores into a mesh based NoC architecture that minimizes the total amount of consumed power by minimizing the total communication among the used cores. Lei and Kumar [7] proposed a two step genetic algorithm for mapping the TG into a mesh based NoC architecture that minimizes the execution time. In the first step, they assumed that all communication delays are the same and selected IP blocks based on the computation delay imposed by the IPs only. In the second step, they used real communication delays.

Murali and De Micheli [8] addressed the problem under the bandwidth constraint with the aim of minimizing communication delay by exploiting the possibility of splitting traffic among various paths. Zhou et al. [10] proposed a multi-objective exploration approach, treating the mapping problem as a two conflicting objective optimization problem that attempts to minimize the average number of hops and achieve a thermal balance. Jena and Sharma [5] addressed the problem of topological mapping of IPs/cores into a mesh-based NoC in two systematic steps using the NSGA-II [1]. The main objective was to obtain a solution that minimizes the energy consumption due to both computational and communicational activities and also minimizes the link bandwidth requirement under some prescribed performance constraints.

10.3 NoC Internal Structure

A NoC platform consisting of architecture and design methodology, which scales from a few dozens to several hundreds or even thousands of resources [6]. As mentioned before, a resource may be a processor core, DSP core, an FPGA block, a dedicated hardware block, mixed signal block, memory block of any kind such as RAM, ROM or CAM or even a combination of these blocks.

A NoC consists of set of *resources* (R) and *switches* (S). Resources and switches are connected by *links*. The pair (R, S) forms a *tile*. The simplest way to connect the available resources and switches is arranging them as a mesh so these are able to communicate with each other by sending messages via an available path. A switch is able to buffer and route messages between resources. Each switch is connected to up to four other neighboring switches through input and output channels. While a channel is sending data another channel can buffer incoming data. Fig. 10.1 shows the architecture of a mesh-based NoC where each resource contains one or more IP blocks (RNI for resource network interface, D for DSP, M for memory, C for cache, P for processor, FP for floating-point unit and Re for reconfigurable block). Besides the mesh topology, there are more complex topologies like *torus*, *hypercube*, 3-*stage clos* and *butterfly*. Note that every resource in the NoC must have an unique identifier and is connected to the network via a switch. It communicates with the switch through the available RNI. Thus, any set of IP blocks can be plugged into the network if its footprint fits into an available resource and if this resource is equipped with an adequate RNI.

10.4 Task Graph and IP Repository Models

In order to formulate the IP mapping problem, it is necessary to introduce a formal definition of an application first. An application can be viewed as a set of tasks that can be executed sequentially or in parallel. It can be represented by a directed graph of tasks, called *task graph*. A *Task Graph* (TG) $G = G(T, D)$ is a directed graph where each node represents a computational module in the application referred to as task $a_i \in T$. Each directed arc $d_{i,j} \in D$, between tasks a_i and a_j, characterizes either

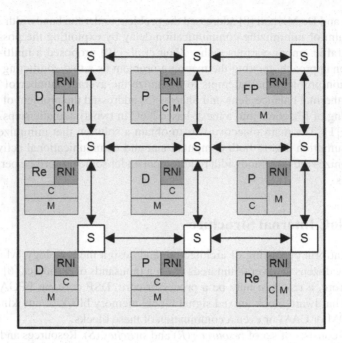

Fig. 10.1 Mesh-based NoC with 9 resources

data or control dependencies. Each task a_i is annotated with relevant information, such as a unique identifier and type of processing element (PE) in the network. Each $d_{i,j}$ is associated with a value $V(d_{i,j})$, which represents the volume of bits exchanged during the communication between tasks a_i and a_j. Once the IP assignment has been completed, each task is associated with an IP identifier. The result of this stage is a graph of IPs representing the PEs responsible of executing the application.

An *Application Characterization Graph* (APG) $G = G(C,A)$ is a directed graph, where each vertex $c_i \in C$ represents a selected IP/core and each directed arc $a_{i,j}$ characterizes the communication process from core c_i to core c_j. Each $a_{i,j}$ can be tagged with IP/application specific information, such as communication rate, communication bandwidth or a weight representing communication cost. A TG is based on application features only while an APG is based on application and IP features, providing us with a much more realistic representation of the an application in runtime on a NoC. In order to be able to bind application and IP features, at least one common feature is required in both of the IP and TG models.

The E3S (0.9) Benchmark Suite [2] contains the characteristics of 17 embedded processors. These processors are characterized by the measured execution times of 47 different type of tasks, power consumption derived from processor data sheets, and additional information, such as die size, price, clock frequency and power consumption during idle state. In addition, E3S contains task graphs of common tasks in auto-industry, networking, telecommunication and office automation. Each one of the nodes of these task graphs is associated with a task type. A task type is a processor instruction or a set of instructions, e.g., FFT, inverse FFT, floating point operation, OSPF/Dijkstra [3], etc. If a given processor is able to execute a given type of instruction, so that processor is a candidate to receive a resource in the NoC structure and would be responsible for the execution of one or more tasks.

Here, we represent TGs using XML code. A TG is divided in three major elements: *taskGraph*, *nodes* and *edges*. Each *node* has two main attributes: an unique identifier (*id*) and a task type (*type*), chosen among the 47 different types of tasks present in the E3S. Each *edge* has four main attributes: an unique identifier (*id*), the *id* of its source node (*src*), the *id* of its target node (*tgt*) and an attribute representing the communication cost imposed (*cost*).

The IP repository is divided into two major elements: the *repository* and the *ips* elements. The *repository* is the IP repository itself. Recall that the repository contains different non general purpose embedded processors and each processor implements up to 47 different types of operations. Not all 47 different types of operations are available in all processors. Each type of operation available in each processor is represented by an *ip* element. Each *ip* is identified by its attribute *id*, which is unique, and by other attributes such as *taskType*, *taskName*, *taskPower*, *taskTime*, *processorID*, *processorName*, *processorWidth*, *processorHeight*, *processorClock*, *processorIdlePower* and *cost*. The common element in TG and IP repository representations is the *type* attribute. Therefore, this element will be used to bind an *ip* to a *node*. The repository contains IPs for digital signal processing, matrix operations, text processing and image manipulation.

These simplified and well-structured representations are easily intelligible, improve information processing and can be universally shared among different NoC design tools.

10.5 Multi-objective Evolution

Optimization problems with *concurrent* and *collaborative* objectives are called Multi-objective Optimization Problems (MOPs). Objectives o_1 and o_2 are said to be collaborative if the optimization of o_1 leads implicitly to the optimization of o_2 while these would be said to be concurrent if the optimization of o_1 leads to the deterioration of o_2. In such problems, all *collaborative* objectives should be grouped and a single objective among those should be used in the optimization process, which achieves also the optimization of all the collaborative objectives in the group. However, concurrent objectives need all to be considered in the process. The best solution for a MOP is the solution with the adequate trade-off between all objectives.

10.5.1 The IP Mapping Problem

The platform-based design methodology for SoC encourages the reuse of components to increase reusability and to reduce the time-to-market of new designs. The designer of NoC-based systems faces two main problems: selecting the adequate set of IPs that optimize the execution of a given application and finding the best physical mapping of these IPs into the NoC structure.

The main objective of the IP assignment stage is to select, from the IP repository, a set of IPs that minimize the NoC consumption of power, area occupied and execution time. At this stage, no information about physical allocation of IPs is available so optimization must be done based on TG and IP information only. So, the result of this step is the set of IPs that maximizes the NoC performance. The TG is then annotated and an APG is produced, wherein each node has an IP associated with it.

Given an application, described by its APG, the problem that we are concerned with in this chapter is to determine how to topologically map the selected IPs onto the network, such that the objectives of interest are optimized. Some of these objectives are: latency requirements, power consumption of communication, total area occupied and thermal behavior. At this stage, a more accurate execution time can be calculated taking into account of the distance between resources and the number of switches and links crossed by a data package along a path. The result of this process should be an optimal allocation of the one of the presecribed IP asssignments, selected in an earlier stage, to execute the application, described by the TG, on the NoC structure.

The search space for a "good" IP mapping for a given application is defined by the possible combinations of IP/tile available in the NoC structure. Assuming that the mesh-based NoC structure has $N \times N$ titles and there are at most N^2 IPs to map, we have a domain size of $N^2!$. Among the huge number of solutions, it is possible to find many equally good solutions. In huge non-continuous search space, deterministic approaches do not deal very well with MOPs. The domination concept introduced by Pareto [9] is necessary to classify solutions. In order to deal with such a big search space and trade-offs offered by different solutions in a reasonable time, a multi-objective evolutionary approach is adopted.

10.5.2   EMO Algorithm

The core of the proposed tool offers the utilization of the well-known and well-tested MOEA: NSGA-II [1]. It adopt the domination concept with a ranking schema for solution classification. The ranking process separates solutions in *Pareto fronts* where each front corresponds to a given rank. Solutions from rank *one*, which is the *Pareto-optimal* front) are equally good and better than any other solution from Pareto fronts of higher ranks.

NSGA-II features a fast and elitist ranking process that minimizes computational complexity and provides a good spread of solutions. The elitist process consists in joining parents and offspring populations and diversity is achieved using the *crowded-comparison operator* [1].

The basic work flow of the algorithm starts with a random population of individuals, where each individual represents a solution. Each individual is associated with a rank. The selection operator is applied to select the parents. The parents pass through crossover and mutation operators to generate an offspring. A new population is created and the process is repeated until the stop criterion is satisfied.

10.5.3 Representation and Genetic Operators

The individual representation is shown in Fig. 10.2–(a). The tile indicates information on the physical location on which a gene is mapped. On a $N \times N$ regular mesh, the tiles are numbered successively from top-left to bottom-right, row by row. The row of the ith tile is given by $\lceil i/N \rceil$, and the corresponding column by i mod N.

The crossover and mutation operators were adapted to the fact that the set of selected IPs can not be changed as we have to adhere to the set of prescribed IP assignments. For this purpose, we propose a crossover operator that acts like a shift register, shifting around a random crossover point and so generating a new solution, but with the same set of IPs. This behavior does not contrast with the biological inspiration of evolutionary algorithms, observing that certain species can reproduce through parthenogenesis, a process in which only one individual is necessary to generate an offspring.

The mutation operator performs an *inner swap mutation*, where each gene receives a random mutation probability, which is compared against the system mutation probability. The genes with mutation probability higher than the system's are swapped with another random gene of the same individual, instead of selecting a random IP from the repository. This way, it is possible to explore the allocation space preserving any optimization done in the IP assignment stage. The crossover and mutation strategies adopted in the IP mapping stage are represented in Fig. 10.2–(b) and Fig. 10.2–(c), respectively.

10.6 Objective Functions

During the evolutionary process, the fitness of the individuals with respect to each one of the selected objectives (i.e. *area*, *time*, and *power*) must be efficiently computed. After a through analysis of all possible design characteristics, we decided that the adequate trade-off can be achieved using only minimization functions of objectives *area*, *execution time* and *power consumption*.

10.6.1 Area

In order to compute the area required by a given mapping, it is necessary to know the area needed for the selected processors and that required by the used links and switches. As a processor can be responsible for more than one task, each APG node must be visited in order to check the processor identification in the appropriate XML element. Grouping the nodes with the same *processorID* attribute allows us to implement this verification. The total number of links and switches can be obtained

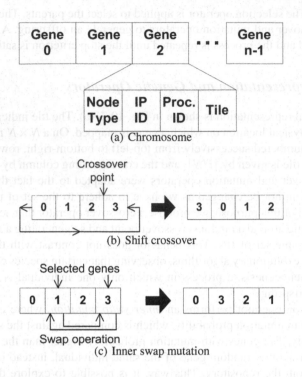

Fig. 10.2 Chromosome and application of the proposed shift crossover and inner swap mutation

through the consideration of all communication paths between exploited tiles. Note that a given IP mapping may not use all the available tiles, links and switches. Also, observe that a portion of a path may be re-used in several communication paths.

In this work, we adopted a fixed route strategy wherein data emanating from tile i is sent first horizontally to the left or right side of the corresponding switch, depending on the target tile position, say j, with respect to i in the NoC mesh, until it reaches the column of tile j, then, it is sent up or down, also depending on the position of tile j with respect to tile i until it reaches the row of the target tile. Each communication path between tiles is stored in the routing table. The number of links in the aforementioned route can be computed as described in Equation 10.1. This is also represents the distance between tiles i and j and called the *Manhattan distance* [7].

$$nLinks(i,j) = |\lceil i/N\rceil - \lceil j/N\rceil| + |i \bmod N - j \bmod N| \qquad (10.1)$$

In the purpose of computing efficiently the area required by all used links and switches, an APG can be associated with a so-called *routing table* whose entries describe appropriately the links and switches necessary to reach a tile from another. The number of hops between tiles along a given path leads to the number of links between those tiles, and incrementing that number by 1 yields the number of traversed switches. The area is computed summing up the areas required by the implementation of all distinct processors, switches and links.

Equation 10.2 describes the computation involved to obtain the total area for the implementation a given IP mapping M, wherein function $Proc(.)$ provides the set of distinct processors used in APG_M and $area_p$ is the required area for processor p, function $Links(.)$ gives the number of distinct links used in APG_M, A_l is the area of any given link and A_s is the area of any given switch.

$$Area(M) = \sum_{p \in Proc(APG_M)} area_p + (A_l + A_s) \times Links(APG_M) + A_s \qquad (10.2)$$

10.6.2 Execution Time

To compute the execution time of a given mapping, we consider the execution time of each task of the critical path, their schedule and the additional time due to data transportation through links and switches along the communication path. The critical path can be found visiting all nodes of all possible paths in the task graph and recording the largest execution time of the so-called critical path. The execution time of each task is defined by the *taskTime* attribute in TG. Links and switches can be counted using the routing table. We identified three situations that can degrade the implementation performance, increasing the execution time of the application:

1. *Parallel tasks mapped into the same tile*: A TG can be viewed as a sequence of horizontal levels, wherein tasks of the same level may be executed in parallel, allowing for a reduction of the overall execution time. When parallel tasks are assigned in the same processor, which also means that these occupy the same tile of the NoC, they cannot be executed in parallel.
2. *Parallel tasks with partially shared communication path*: When a task in a tile must send data to supposedly parallel tasks in different tiles through the same *initial* link, data to both tiles cannot be sent at the same time.
3. *Parallel tasks with common target using the same communication path*: When several tasks need to send data to a common target task, one or more shared links along the partially shared path would be needed simultaneously. The data from both tasks must then be pipelined and so will not arrive at the same time to the target task.

Equation 10.3 is computed using a recursive function that implements a depth-first search, wherein function $Paths(.)$ provides all possible paths of a given APG and $t_0(a)$ is the required time for task a. After finding the including the total execution time of the tasks that are traversed by the critical path, the time of parallel tasks executed in the same processor need to be accumulated too. This is done by function $SameProcSameLevel(.)$. The delay due to data pipelining for tasks on the same level is added by $SameSourceCommonPath(.)$. Last but not least, the delay due to pipelining data that are emanating at the same time from several distinct tasks yet for the same target task is accounted for by function $DiffSrcSameTgt(.)$.

$$Time(M) = \max_{r \in Paths(APG_M)} \left(\sum_{a \in r} t_0(a) + \sum_{i \in \{1,2,3\}} t_i(r) \right) \tag{10.3}$$

Function $t_1 - SameProcSameLevel(.)$ compares tasks of a given same level that are implemented by the same processor and returns the additional delay introduced in the execution of those tasks. Algorithm 10.1 shows how function $SameProcLevel(.)$, that uses information from path r, application task graph and its corresponding characterization graph to compute the delay in question.

Algorithm 10.1. SameProcSameLevel$(r) - t_1$

```
 1: time := 0
 2: for all a ∈ r do
 3:    for all n ∈ T do
 4:       if T.level(a) = TG.level(n) then
 5:          if APG.processor(a) = APG.processor(n) then
 6:             time := time + n.taskTime
 7:          end if
 8:       end if
 9:    end for
10: end for
11: return time
```

Function $t_2 - SameSourceCommonPath(.)$ computes the additional time due to parallel tasks that have data dependencies on tasks mapped in the same source tile and yet these share a common initial link in the communication path. Algorithm 10.2 shows the details of the delay computation using information from path r, application task graph and its corresponding characterization graph. In that algorithm $TG.targets(a)$ yields the list of all possible target tasks of task a, $APG.initPath(src, tgt)$ returns the initial link of the communication path between tiles src and tgt and $penalty$ represents a time duration needed to data to cross safely from one switch to one of its neighbors. This penalty is added every time the initial link is shared.

Function $t_3 - DiffSrcSameTgt(.)$ computes the additional time due to the fact that parallel tasks producing data for the same target task need to use simultaneously at least a common link along the communication path. Algorithm 10.3 shows the details of the delay computation using information from path r, application task graph and its corresponding characterization graph. In that algorithm, $APG.Path(src, tgt)$ is the ordered list of all links crossed from task src to task tgt and $penalty$ has the same semantic as in the Algorithm 10.2.

10.6.3 Power Consumption

The total power consumption of an application NoC-based implementation consists of the power consumption of the processors while processing the computation

---

Algorithm 10.2. SameSrcCommonPath(r) – t_2

1: $penalty := 0$
2: **for all** $a \in r$ **do**
3:      **if** $TG.targets(a) > 1$ **then**
4:          **for all** $n \in TG.targets(a)$ **do**
5:              **for all** $n' \in TG.targets(a) \mid n' \neq n$ **do**
6:                  $w = APG.initPath(a,n)$;
7:                  $w' = APG.initPath(a,n')$;
8:                  **if** $w = w'$ **then**
9:                      $penalty := penalty + 1$
10:                 **end if**
11:             **end for**
12:         **end for**
13:     **end if**
14: **end for**
15: **return** $penalty$

Algorithm 10.3. DiffSrcSameTgt(r) – t_3

1: $penalty := 0$
2: **for all** $a \in r$ **do**
3:      **for all** $a' \in r \mid a' \neq t$ **do**
4:          **if** $TG.level(a) = TG.level(a')$ **then**
5:              **for all** $n \in TG.targets(a)$ **do**
6:                  **for all** $n' \in TG.targets(a')$ **do**
7:                      **if** $n = n'$ **then**
8:                          $w := APG.Path(a,n)$;
9:                          $w' := APG.Path(a',n')$;
10:                         **for** $i = 0$ to $min(w.length, w'.length)$ **do**
11:                             **if** $w(i) = w'(i)$ **then**
12:                                 $penalty := penalty + 1$
13:                             **end if**
14:                         **end for**
15:                     **end if**
16:                 **end for**
17:             **end for**
18:         **end if**
19:     **end for**
20: **end for**
21: **return** $penalty$

---

performed by each IP and that due to the data transportation between the tiles. The
former can be computed summing up attribute *taskPower* of all nodes of the APG
and the latter is the power consumption due to communication between the application tasks through links and switches. The power consumption due to the computational activity is simply obtained summing up atribute *taskPower* of all nodes in the
APG and is as described in Equation 10.4.

$$Power_p(M) = \sum_{a \in APG_M} power_a \qquad (10.4)$$

An energy model for one bit consumption is used to compute the total energy consumption for the whole communication involved during the execution of an application on the NoC platform. The bit energy (E_{bit}), energy consumed when a data of one bit is transported from one tile to any of its neighboring tiles, can be obtained as in Equation 10.5:

$$E_{bit} = E_{S_{bit}} + E_{L_{bit}} \qquad (10.5)$$

wherein $E_{S_{bit}}$ and $E_{L_{bit}}$ represent the energy consumed by the switch and link tying the two neighboring tiles, respectively [4].

The total power consumption of sending one bit of data from tile i to tile j can be calculated considering the number of switches and links the bit passes through on its way along the path, as shown in Equation 10.6.

$$E_{bit}^{i,j} = nLinks(i,j) \times E_{L_{bit}} + (nLinks(i,j)+1) \times E_{S_{bit}} \qquad (10.6)$$

wherein function $nLinks(.)$ provides the number of traversed links (and switches too) considering the routing strategy used in this work and described earlier in this section. The function is is defined in Equation 10.1.

Recall that the application TG gives the communication volume ($V(a,a')$) in terms of number of bits sent from the task a to task a' passing through a direct arc $d_{a,a'}$. Assuming that the tasks a and a' have been mapped onto tiles i and j respectively, the communication volume of bits between tiles i and j is then $V(i,j) = V(d_{t,t'})$. The communication between tiles i and j may consist of a single link $l_{i,j}$ or by a sequence of $m > 1$ links $l_{i,x_0}, l_{x_0,x_1}, l_{x_1,x_2}, \ldots, l_{x_{m-1},j}$.

The total network communication power consumption for a given mapping M is given in Equation 10.7, wherein $Targets_a$ provides all tasks that have a direct dependency on data resulted from task a and $Tile_a$ yields the tile number into which task a is mapped.

$$Power_c(M) = \sum_{\substack{a \in APG_M, \\ \forall a' \in Targets_a}} V(d_{a,a'}) \times E_{bit}^{Tile_a, Tile_{a'}} \qquad (10.7)$$

10.7 Results

First of all, the implementation of the algorithm was validated using mathematical known MOPs and the results were compared with the original results that were obtained by Deb to validate NSGA-II [1]. The simulation converged to the true Pareto-front. For NoC optimization, only the individual representation and the objective functions were changed, keeping the ranking, selection, crossover and mutation operators unchanged. Different TGs generated with TGFF [2] and from E3S, with sequential and parallel tasks, were used.

Many simulations were performed to find out the setting up of the parameters used in NSGA-II for solving the IP mapping problem. The results of these simulation allowed us to set the population size to 600, mutation probability to 0.01, crossover probability to 0.8 and tournament size to 50 and run the algorithm of 100 generations. The application, represented as a TG in Fig. 10.3, was generated with TGFF [2]. Note that this TG presents four levels of parallelism.

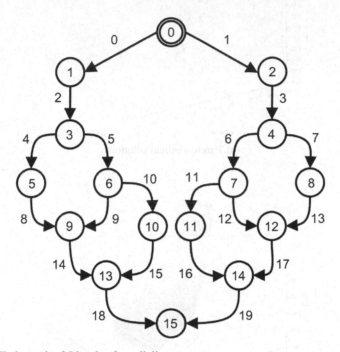

Fig. 10.3 Task graph of 5 levels of parallelism

Analyzing the results obtained from the first simulations, we observed that in order to achieve the best trade-off, the system allocated many tasks for the same processor, which reduces area and execution time but generates *hot spots* [10]. A hot spot is an area of high activity within a silicon chip. Hot spots can damage a silicon chip and increases power consumption because of Avalanche Effect. In order to avoid the formation of hot spots, a *maximum tasks per processor* constraint was imposed in the evolutionary process. This parameter is decided by the NoC designer based on some extra physical characteristics. We adopted a maximum of 2 tasks per processor. Figure 10.4–(a) shows the Pareto-front discrete points. Figure 10.4–(b) shows the Pareto-front formed by the Pareto-optimal solutions. Note that many solutions have very close objectives values. The IP assignment of the TG represented in Fig. 10.3 was able to discover 97 distinct optimal IP assignments. From those 97 distinct of IP assignments, 142 optimal mappings were generated.

Fig. 10.5–(a) represents the *time × area* trade-off, Fig. 10.5–(b) depicts the *power × time* trade-off and Fig. 10.5–(c) plots the *power × area* trade-off. As we

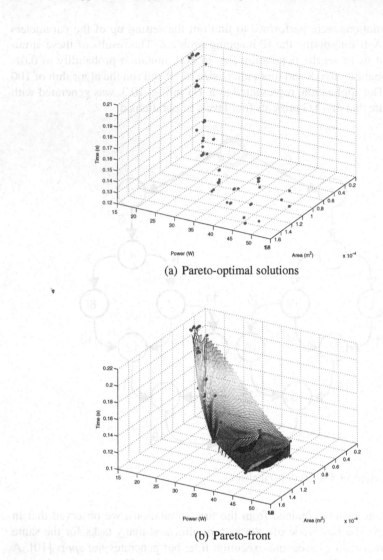

(a) Pareto-optimal solutions

(b) Pareto-front

Fig. 10.4 Pareto-optimal solutions and Pareto-front of the 142 optimal IP mappings obtained for the task graph of Fig. 10.3

can observe, comparing the dots against the line of interpolation, the trade-off between time and area and between power and time is not so linear as the trade-off between power and area. Fig. 10.5–(a) shows that solutions that require more area tend to spend less execution time because of the better distribution of the tasks allowing for more parallelism to occur. Fig. 10.5–(b) shows that solutions that spend less time of execution tend to consume more power because of IP's features, such as higher clock frequency, and physical effects like intensive inner-electrons activity.

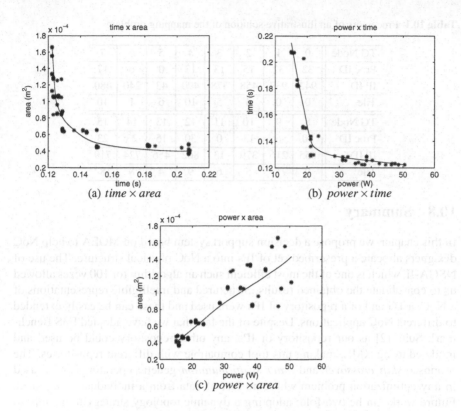

Fig. 10.5 Trade-offs representation of the 142 IP mappings for the task graph of Fig. 10.3

Fig. 10.5–(c) shows a linear relation between power consumption and area. Those values and units are based on E3S Benchmark Suite [2].

For a TG of 16 tasks, a 4 × 4 mesh-based NoC is the maximal physical structure necessary to accommodate the corresponding application. The obtained solutions showed that no solution used more than ten resources to map all tasks. The unused 6 tiles may denote a waste of hardware resources, which consequently lead to the conclusion that either the geometry of the NoC is not suitable for this application or the mesh-based NoC is not the ideal topology for its implementation.

As a specific mapping example, we detail one of the solutions, which seems to be a moderate solution with respect to every considered objectives. Table 10.1 specifies the processors used in the solution. We can observe that all parallel tasks were allocated in the distinct processors, which reduces execution time. The number of processors were minimized based on the optimization of the objectives of interest and this minimization was controlled by the maximum tasks per processor constraint to avoid hot spots [10]. The processors were allocated in such way to avoid delay of communication due to links and switches disputed by more than one resource at the same time.

Table 10.1 Processors of an illustrative solution of the mapping problem

TG Node	0	1	2	3	4	5	6	7
Proc ID	32	32	15	13	17	0	6	17
IP ID	942	937	458	378	490	43	240	480
Tile	0	0	4	5	10	6	1	10
TG Node	8	9	10	11	12	13	14	15
Proc ID	30	6	13	0	30	15	23	23
IP ID	855	216	379	13	862	456	724	719
Tile	9	1	5	6	9	4	8	8

10.8 Summary

In this chapter, we propose a decision support system based on MOEA to help NoC designers allocate a prescribed set of IPs into a NoC physical structure. The use of NSGA-II, which is one of the most efficient such an algorithm for 100 vezes allowed us to consolidate the obtained results. Structured and intelligible representations of a NoC, a TG and of a repository of IPs were used and these can be easily extended to different NoC applications. Despite of the fact that we have adopted E3S Benchmark Suite [2] as our repository of IPs, any other repository could be used and modeled using XML, making this tool compatible with different repositories. The proposed *shift crossover* and *inner swap mutation* genetic operators can be used in any optimization problem where no lost of data from a individual is accepted. Future work can be two-fold: adopting a dynamic topology strategy to attempt to evolve the most adequate topology for a given application and exploring the use of different objectives based on different repositories.

References

1. Deb, K., Pratap, A., Agarwal, S., Meyarivan, T.: A fast and elitist multiobjective genetic algorithm: NSGA-II. IEEE-EC 6, 182–197 (2002)
2. Dick, R.P., Rhodes, D.L., Wolf, W.: TGFF: Task Graphs For Free. In: Proceedings of the 6th International Workshop on Hardware/Software Co-design, pp. 97–101. IEEE Computer Society, Seattle (1998)
3. Dijkstra, E.W.: A note on two problems in connexion with graphs. Numerische Mathematik 1, 269–271 (1959)
4. Hu, J., Marculescu, R.: Energy-aware mapping for tile-based NoC architectures under performance constraints. In: ASPDAC: Proceedings of the 2003 Conference on Asia South Pacific Design Automation, pp. 233–239. ACM, New York (2003)
5. Jena, R.K., Sharma, G.K.: A multi-objective evolutionary algorithm based optimization model for network-on-chip synthesis. In: ITNG, pp. 977–982. IEEE Computer Society (2007)
6. Kumar, S., Jantsch, A., Millberg, M., Öberg, J., Soininen, J.-P., Forsell, M., Tiensyrjä, K., Hemani, A.: A network on chip architecture and design methodology. In: ISVLSI, pp. 117–124. IEEE Computer Society (2002)

7. Lei, T., Kumar, S.: A two-step genetic algorithm for mapping task graphs to a network on chip architecture. In: DSD, pp. 180–189. IEEE Computer Society (2003)
8. Murali, S., Micheli, G.D.: Bandwidth-constrained mapping of cores onto NoC architectures. In: DATE, pp. 896–903. IEEE Computer Society (2004)
9. Pareto, V.: Cours D'Economie Politique. F. Rouge, Lausanne (1896)
10. Zhou, W., Zhang, Y., Mao, Z.: Pareto based multi-objective mapping IP cores onto NoC architectures. In: APCCAS, pp. 331–334. IEEE (2006)

Chapter 11
Routing in Network-on-Chips Using Ant Colony Optimization*

Abstract. Networks-on-Chip (NoC) have been used as an interesting option in design of communication infrastructures for embedded systems, providing a scalable structure and balancing the communication between cores. Because several data packets can be transmitted simultaneously through the network, an efficient routing strategy must be used in order to avoid congestion delays. In this chapter, ant colony algorithms were used to find and optimize routes in a mesh-based NoC, where several randomly generated applications have been mapped. The routing optimization is driven by the minimization of total latency in packets transmission between tasks. The simulation results show the effectiveness of the ant colony inspired routing by comparing it with general purpose algorithms for deadlock free routing.

11.1 Introduction

A System-on-Chip (SoC) is an integrated circuit composed by a full computer system. SoCs contains, within the same package, processors, memory, input-output controllers and specific application devices. This block structure follows a design methodology based on intellectual property (IP) cores. Components designed for a specific project can be reused in other SoCs, reducing design time. Thus, under an extremely simplified view, to increase the number of tasks performed by the SoC, just add more IP cores with different features.

The increase of SoCs scale raises new design challenges. Among them is communication between IP cores. The blocks of a SoC are interconnected by a communication infrastructure, such as buses or point-to-point links. However, each of these models have their limitations. Shared buses can cause high delays if multiple blocks need to transmit data simultaneously. This does not happen in point-to-point architectures. In turn, the communication structure need to be redesigned for each new system. For many SoC designs, it is desirable to use a framework scalable as buses and fast as point-to-point links. An architecture that includes these two features are the NoCs, *Networks-on-Chip* [1].

---

* This chapter was developed in collaboration with Luneque Del Rio de Souza e Silva Júnior.

N. Nedjah and L. de Macedo Mourelle, *Hardware for Soft Computing and Soft Computing for Hardware*, Studies in Computational Intelligence 529,
DOI: 10.1007/978-3-319-03110-1_11, © Springer International Publishing Switzerland 2014

In an NoC architecture, *switches* are interconnected by point-to-point links, thus describing a network topology. An example of network topology is the mesh shown in Fig. 11.1. The switches are also connected to the IP cores that constitute the system, also called *resources*. Switches exchange information in the form of messages and packages. The information generated by a resource is divided into smaller parts and sent over the network. These packages are organized in the destination switch and then delivered to resource. This operation is similar to that performed by computer networks. The structure formed by a switch and a resource is called a *network node*. NoCs can be used in the implementation of multi-processors systems-on-chip (MPSoCs) for running applications with high level of parallelism [23].

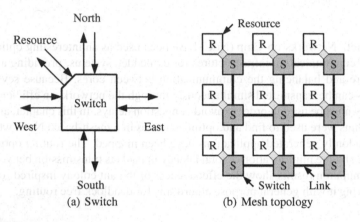

(a) Switch   (b) Mesh topology

Fig. 11.1 Network-on-chip architecture

In the design of NoC-based systems, the communication infrastructure can be imported as a single configurable IP block. However, many are the ways to connect network and resources, in order to achieve the desired application. To assist the designer, computational tools for project assistance, or EDAs (Electronic Design Automation), are used [20]. The purpose of EDAs is to optimize intermediate stages of SoC and NoCs project, in order to obtain a more efficient design implementation [15].

In general, NoCs are developed to perform a specific application. This application can be described initially as a software that must be embedded in hardware. The EDA tool must be able to use information about the desired application (at a high level of abstraction) and, through successive stages of optimization, implement a solution that meets the design specifications, which may include hardware area, power consumption and time of execution. This optimization may include several steps, such as *task allocation* [5], *IP mapping* [24] and *static routing*. The Fig. 11.2 shows in a simplified way the flowchart of a SoC design based on network-on-chip.

The process of IP allocation consists in associating each task (or set of tasks) to an appropriate IP block within a set of IPs or repository capable of performing such

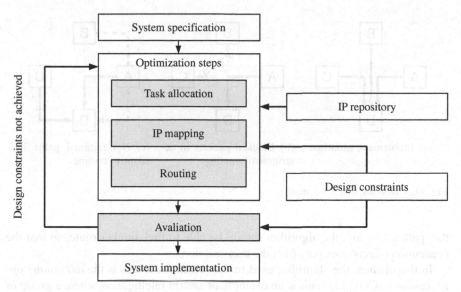

Fig. 11.2 Typical embedded system design flow for NoC platform

a task. The mapping of an application consists in associating the set of IPs resulting from the allocation to each node in communication infrastructure - in this case, the NoC. In other words, is spatially defined where each feature will be implemented, i.e., where in network each IP core is connected. Routing, in turn, defines which switches will be used for communication between cores.

Delays in communication may occur in congestion situations, when multiple packets could be transmitted using the same switch at the same time. If the routing algorithm adopted in the NoCs design is deterministic, the selection of the packet path from the source to the destination switch will not consider the load of intermediate switches - those between the source and destination switch. If these switches are under a heavy traffic, a given packet can only be transmitted after the end of congestion. This occurs even if other switches, not selected for routing, are free for transmission. On the other hand, adaptive routing algorithms can be used in order to avoid network congestion. These algorithms use not only the position of origin and destination nodes, but also the actual load condition of the network to calculate the route. When you find a region of network in use, the routing can set the package to follow another path. This congestion-free path may, however, be not minimal. These two situations are shown in Fig. 11.3.

In order to overcome the congestion problem, this chapter proposes a route optimization step in the design of NoCs, or more precisely, an adaptive and static routing. In this routing, a network model provides the communication patterns required for application execution. The calculation of routes is accomplished by an optimization algorithm to minimize the communication time. The search is always for a shortest path between origin and destination. If the intermediate switches of

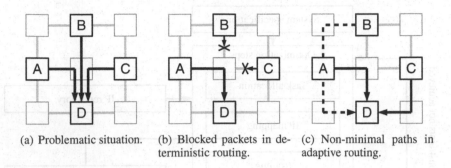

(a) Problematic situation.   (b) Blocked packets in de-   (c) Non-minimal paths in terministic routing.   adaptive routing.

Fig. 11.3 Routing in a 3 × 3 mesh

this path are in use, the algorithm should be able to find another route, so that the contention effects does not affect the transmission.

In this chapter, the algorithm used in the search for routes is the *ant colony optimization* (ACO) [12]. This is an example of swarm intelligence, where a group of individuals work together to find a solution to a given problem. We compared the results of the network using the proposed routing algorithms and literature widely adopted routing algorithms.

The reminder of this chapter is organized as follows. In Section 11.2 we review the related work in routing algorithms. The specification of the simulated network is shown in Section 11.3. In Section 11.4, we do an overview on ACO meta-heuristic. The proposed routing is presented in Section 11.5. A brief description of applications and mapping is shown in Section 11.6. Simulation results are presented in Section 11.7. The chapter closes with a conclusion and the description of future work in Section 11.8.

11.2 Routing in Communication Networks

There are several works that study the efficient routing in parallel and distributed computing. For a broader reference, [26] presents a survey of routing techniques for direct networks.

Many of the techniques used for routing in NoCs, such as the *XY algorithm*, were originally developed for computer networks and multiprocessor systems. The XY algorithm is a routing technique widely used in 2D mesh networks with wormhole switching, such as the Intel Touchstone DELTA [19], the Intel Paragon [16], the Symult 2010 [27] and the Caltech MOASIC [28]. It works by sending packets over the network first horizontally (X dimension), then vertically (Y dimension). This idea can be expanded to a larger number of dimensions, being known as such DOR (dimension order routing) [26]. In the context of NoCs, XY routing proves efficient due to its simplicity of implementation and because it is deadlock-free. Works that made use of this algorithm include the HERMES network [22] and the SoCIN network [29].

Glass and Ni have proposed the so-called *Turn Model* for adaptive, livelock and deadlock free algorithms [17]. A turn is a change of 90° in the direction of packet transmission. The main idea of this model is to restrict the amount of turns that a packet route can go through in order to avoid the formation of cycles that cause deadlocks. Following this concept, three routing algorithms were proposed by Glass and Ni: the *West-First*, the *North-last* and *Negative-First*. A related approach is the *Odd-Even* turn model [3] for designing partially adaptive deadlock-free routing algorithms. Unlike the turn model, which relies on prohibiting certain turns in order to avoid deadlock, this model restricts the locations where some types of turns can be taken. As a result, the degree of routing adaptiveness provided is more even for different source-destination pairs.

The work of Jose Duato has addressed the mathematical foundations of routing algorithms. His main interests have been in the area of adaptive routing algorithms for multicomputer networks. Most of the concepts are directly applicable to NoC. In [13], the theoretical foundation for deadlock-free adaptive routing in wormhole networks is given.

11.3 Network Specification

The network model in this work uses switches with five communication ports. Four ports are responsible for communication with neighboring switches and one is for local communication with the resource. The switches are considered bufferless using no virtual channels. The network topology is a two dimension mesh, as shown in Fig 11.1. The switching technique adopted was the wormhole. In this method, packets are divided into smaller units called *flits* (flow-units). It is assumed that each communication channel has a width of a flit. The transmission of flits is performed in a pipeline way, as seen in Fig. 11.4.

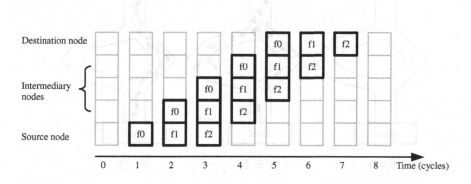

Fig. 11.4 Transmission of 3 flits in wormhole switching

The latency of a packet sent trough the network in wormhole switching is given by Equation 11.1, where t_{flit} is the transmission time of a flit in a channel, D is the number of switches in a path, L is the total length of a packet (in bits), W is the length of a channel, and L_{delay} is the number of bits that would have been transmitted in a period of congestion.

$$T_{packet} = t_{flit} \cdot \left(D + \left\lceil \frac{L}{W} \right\rceil + \left\lceil \frac{L_{delay}}{W} \right\rceil \right) \tag{11.1}$$

11.4   Ant Colony Optimization

Ant algorithms, also known as Ant Colony Optimization (ACO) [12], are a class of heuristics search algorithms, that have been successfully applied to solving NP hard problems [2]. Ant algorithms are biologically inspired in the behavior of colonies of ants, and in particular how they forage for food. One of the main ideas behind this approach is that the ants can communicate with one another through indirect means by making modifications to the concentration of highly volatile chemicals called *pheromones* in their neighbor environment. As it has been shown [18], indirect communication among ants via pheromone trails enables them to find shortest paths between their nest and food sources. The most emphatic and best known example of the use of pheromones by ants is the double bridge experiment. An ant nest is connected to a food source by two bridges with different lengths. This configuration is shown in Fig. 11.5. Initially, ants choose equally both ways. However, opting ants for shorter path are able to go back to the food supply before the ants that follow the long way. Thus, also the concentration of pheromone on the shortest path will be greater from the moment the ants complete the round trip. Consequently, next ants will be more attracted by bridge with more pheromone, i.e., the shorter path.

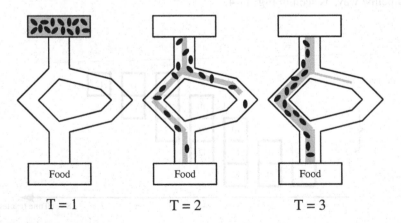

Fig. 11.5 Pheromone concentration in the double bridge experiment

This capability of real ant colonies has inspired the definition of artificial colonies, that can find approximate solutions to hard combinatorial optimization problems. The main ideas of ACO are the use of:

- repeated simulations carried out by a population of artificial agents called "ants" to generate new solutions to the problem;
- stochastic local search to build the solutions in an incremental way;
- information collected during past simulations (artificial pheromones) to direct future search for better solutions.

Several ant algorithms make use of the structure shown in the Algorithm 11.1, the ACO meta-heuristics [10].

Algorithm 11.1. ACO meta-heuristics

1: initialize parameters and pheromone trails;
2: **while** termination condition not met **do**
3:    construct ant solutions;
4:    local search (optional);
5:    update pheromone trails;
6: **end while**;

In the artificial ant colony approach, each ant builds a solution by using two types of information locally accessible: problem-specific information, and information added by ants during previous iterations of the algorithm. In fact, while building a solution, each ant collects information on the problem characteristics and on its own performance, and uses this information to modify the representation of problem, as seen locally by the other ants. The representation of the problem is modified in such a way that information contained in past good solutions can be exploited to build new and hopefully better ones. This form of indirect communication mediated by the environment is called *stigmergy*, and is typical in social insects.

11.5 ACO-Based Routing

The Ant Colony Optimization, with the ability to search for paths, emerging as a powerful solution for routing problems. Thus, this chapter presents the use of the ACO meta-heuristic in the construction of routing algorithms. Two models of static routing for NoCs are proposed. These algorithms were called REAS (routing based on EAS [12]) and RACS (routing based on ACS [11]). Both algorithms search paths in an *architecture characterization graph* that represents the network 2D mesh topology. These algorithms make use of multiple ant colonies, where each colony is responsible for searching the route of a package. In this approach, each colony has its own pheromone and ants. However, the colonies must exchange information in order to minimize the latency of their respective packages. Thus, the route found by an ant from a given colony is visible to the ants from other colonies, because these packets are being transmitted simultaneously and in the same network. In the proposed algorithms, ants in a network node knows only two things. The first is the pheromone concentration in the surrounding nodes. The second is the load on a node, the waiting time in each of the four possible transmission directions.

11.5.1 REAS Algorithm

The *Elitist Ant System* is directly inspired by the *Ant System*, the first or ant algorithms [12]. The EAS is characterized mainly by the use of *elitism*, in order to differentiate the best ants. A simplified pseudo-code of REAS is shown in Algorithm 11.2.

Algorithm 11.2. REAS algorithm

Require: network parameters;
Require: EAS parameters;
Require: packets parameters;
  1: **while** total of cycles **do**
  2:     **for** $k = 1 \rightarrow number\ of\ ants$ **do**
  3:       **for** $g = 1 \rightarrow number\ of\ packets$ **do**
  4:         **while** $node_{actual} \neq node_{destination}$ **do**
  5:           $Ant_{k,g}$ select the $node_{next}$;
  6:           calculates the load of $Ant_{k,g}$ in $node_{actual}$;
  7:           $node_{actual} \leftarrow node_{next}$
  8:         **end while**
  9:         calculates $Ant_{k,g}$ pheromone;
 10:       **end for**
 11:       calculates the elitist pheromone;
 12:       accumulate the pheromone of ants in k iteration;
 13:     **end for**
 14:     update the global pheromone;
 15: **end while**
 16: **return** best solution;

In the REAS algorithm, ants build paths through the network selecting the next node with base in Equation 11.2, where p_{ij}^k is the probability of the ant k go from the node i to the node j.

$$p_{ij}^k(t) = \begin{cases} \dfrac{\tau_j(t)^\alpha \cdot \eta_{ij}{}^\beta}{\sum\limits_{k \in allowed_k} \tau_k(t)^\alpha \cdot \eta_{ik}{}^\beta} & \text{if } j \in allowed_k \\[4mm] 0 & \text{otherwise} \end{cases} \tag{11.2}$$

The probability of selecting a particular direction is a function of pheromone concentration and network load in this direction. These two parameters are weighted by their importance constant α and β. The network load is used indirectly by η_{ij}, defined by:

$$\eta_{ij} = \frac{1}{C_{ij}} \tag{11.3}$$

where C_{ij} is the load in transmission from i to j.

At the end of each iterative cycle, the pheromone of all colonies is updated according to Equation 11.4. Part of the pheromone of the previous iteration is reduced by evaporation rate ρ, and then reinforced by the contribution of all m ants in the current cycle. The pheromone also receives the reinforcement of elitist ants: those that achieve the best solutions deposit their pheromone in every cycle, directing the search in subsequent cycles.

$$\tau_{t+1} = (1-\rho)\cdot\tau_t + \sum_{k=1}^{m}\Delta\tau^k + \tau_{elite} \tag{11.4}$$

The pheromone in the path find by a single ant k is defined by:

$$\Delta\tau^k = \frac{Q}{L_k} \tag{11.5}$$

where Q is a constant and L_k represents the total latency of the solution. It is easy to see that the ants with the worst results provide a smaller amount of pheromone.

11.5.2 RACS Algorithm

The second ant algorithm used in this work is described below. The RACS is very similar to REAS, with the same structure of *multiple colonies* being used. The algorithm on which the RACS was inspired, called *Ant Colony System* [11], differs from others ant algorithms by:

- the selection method of next nodes in solutions building;
- the use of a different pheromone update.

Because these two mechanisms, ACS improves over AS by increasing the importance of exploitation of information collected by previous ants with respect to exploration of the search space. The pseudo-code of RACS algorithm is shown in Algorithm 11.3.

Thus, the RACS uses the so-called *pseudo-random proportional* rule.

$$j = \begin{cases} argmax_{j\in[1,4]}\left\{\tau_j\cdot\eta_{ij}^\beta\right\} & \text{if } q \leq q_0, \\ S & \text{otherwise} \end{cases} \tag{11.6}$$

As shown in Equation 11.6, the probability for an ant to move from node i to node j depends on a random variable q, uniformly distributed over $[0,1]$, and a parameter q_0. If $q \leq q_0$, then the next node is directly selected by $argmax_{j\in[1,4]}\{\tau_j\cdot\eta_{ij}^\beta\}$, i.e., the direction with the largest value of $\tau_j\cdot\eta_{ij}^\beta$. Otherwise, the next node is defined by S, that uses a selection method similar to that employed by EAS (Equation 11.2).

Algorithm 11.3. RACS algorithm

Require: network parameters;
Require: ACS parameters;
Require: packets parameters;
 1: **while** total of cycles **do**
 2:   **for** $k = 1 \rightarrow$ *number of ants* **do**
 3:     **for** $g = 1 \rightarrow$ *number of packets* **do**
 4:       **while** $node_{actual} \neq node_{destination}$ **do**
 5:         $Ant_{k,g}$ select the $node_{next}$;
 6:         calculates the load of $Ant_{k,g}$ in $node_{actual}$;
 7:         update the local pheromone in $node_{actual}$;
 8:         $node_{actual} \leftarrow node_{next}$
 9:       **end while**
10:       calculates $Ant_{k,g}$ pheromone;
11:     **end for**
12:     **if** solution of ants in k iteration is the best **then**
13:       $\tau_{best} \leftarrow$ pheromone of ants in k iteration;
14:     **end if**
15:   **end for**
16:   update the global pheromone with τ_{best};
17: **end while**
18: **return** best solution;

The RACS algorithm also uses a double pheromone update. The *offline update* is applied at the end of each iteration only by the *best-so-far* ant.

$$\tau_{t+1}^j = \begin{cases} (1-\rho) \cdot \tau_t^j + \rho \cdot \Delta\tau_j & \text{if } j \text{ belongs to best path} \\ \tau_t^j & \text{otherwise} \end{cases} \quad (11.7)$$

The offline update is given by Equation 11.7, where $\Delta\tau_j$ is the reinforcement of the best ant pheromone. As said, the offline update perform a strong elitist strategy. The best ant can be the iteration-best ant, that is, the best in the current iteration, or the global-best ant, that is, the ant that made the best tour from the start of the trial.

The *local update* is performed by all ants in each step of construction of a solution.

$$\tau_{t+1} = (1-\rho) \cdot \tau_t + \rho \cdot \tau_0 \quad (11.8)$$

This local update is defined by Equation 11.8, where ρ is the evaporation constant, and τ_0 is the initial pheromone at each node. In practice, ACS ants consume some of the pheromone trail on the nodes they visit. This has the effect of decreasing the probability that a same path is used by all the ants (i.e., it favors exploration, counterbalancing this way the above-mentioned modifications that strongly favor exploitation of the collected knowledge about the problem).

11.6 Applications in NoC

In general, NoCs are developed to perform a specific application. This application can be described initially as a software that must be embedded in hardware. The EDA tool must adjust characteristics of NoC and application so that the execution is more efficient.

11.6.1 Task Graphs

Every application, in any type of computer system, can be described by a *task graph*. This is a data structure in which the application is divided into blocks responsible for specific tasks. These blocks, in turn, exchange information in order to complete the application execution. The number of blocks may be higher or lower depending on the level of abstraction adopted. Thus, the task graph is denoted by $GT = G(T,D)$, an acyclic and weighted directed graph. Each node of T is a task, or an application processing module. In general, an operation is a well defined task, as a mathematical calculation or a data encoding. Each arc of D characterizes the data dependencies between two tasks.

11.6.2 Random Mapping

As mentioned in Section 11.1, an EDA tool can perform certain processes, such as allocation, mapping and routing of applications in NoCs. In the allocation process, cores are selected in an IP repository, and then associated with each application task. In turn, the mapping deals with how the IPs are spatially distributed in the NoC topology. Both processes are intended to optimize some characteristic of system, like execution time, silicon area and power consumption [25, 6].

In this chapter, the routing is emphasized. Thus, it is considered that an allocation step was previously performed, and the resource specifications are already available for use. In the mapping step, we employed a simple *random mapping* process. In the random mapping, each node of the application characterization graph is associated to a node of the architecture characterization graph. The way in which this association is made is random: a node of the application graph selects from a list of a node of the architecture graph; the chosen position is removed from the list and the process repeats until all nodes of the application graph have a defined position in the architecture graph.

The mapping also defines the number of nodes in NoC. The number N of nodes in a mesh must be sufficient to map an application with P tasks. Thus, because it's a 2D mesh, the relationship between N and P is defined by (11.9).

$$N = \left\lceil \sqrt{P} \right\rceil^2 \tag{11.9}$$

The Fig. 11.6 illustrates the mapping process. The tasks of the graph are associated with five nodes in a 3×3 network.

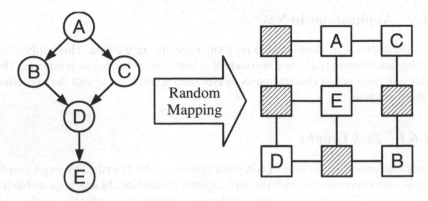

Fig. 11.6 Application *task graph* mapped randomly in a NoC

11.7   Evaluation Experiments and Results

A cycle-accurate network simulator was implemented in Matlab. It supports 2D
mesh networks with wormhole switching. To evaluate the performance of the proposed methods, networks were simulated with four different routing algorithms:
REAS, RACS, XY and Odd-Even (OE). The time unit adopted is the *simulator
cycle*, where one cycle is the transmission time of one flit.

   All algorithms were executed with Matlab Version 7.7.0.471 (R008b). The simulations were performed on PCs with Intel Core i7 950 3GHz, 8Gb RAM and Mi-
crosoft Windows 7 Home Premium operating system.

11.7.1   Tests with Synthetic Traffic Patterns

The network was simulated with size of 5 × 5, a square of 25 nodes. The set of simulation tests were performed varying the network routing algorithm, the pattern of
traffic generation, the rate of injection and the number of packets. These parameters
are shown in Table 11.1.

Table 11.1 Simulation parameters

Routing algorithms	REAS, RACS, XY, OE
Traffic pattern	Uniform, Hots-pot, Local, Complement, Trans. 1, Trans. 2
Injection rate	10%, 20%, 30%, 40%, 50%, 60%, 70%, 80%, 90%, 100%
Number of packets	10, 20, 30, 40, 50, 60, 70, 80, 90, 100

   According to [14], the evaluation of interconnection networks requires the definition of representative workload models. This is a difficult task because the behavior
of the network may differ considerably from one architecture to another and from

one application to another. Moreover, in general, performance is more heavily affected by traffic conditions than by design parameters. Up to now, there has been no agreement on a set of standard traces that could be used for network evaluation. Most performance analysis used synthetic workloads with different characteristics. These models can be used in the absence of more detailed information about the applications. Workload models are basically defined by three parameters: *distribution of destinations*, *injection rate*, and *message length*.

11.7.1.1   Packet Distribution

The source-destination pairs are generated following six different distribution patterns, as shown in Fig. 11.7. These patterns are based on models widely used in the evaluation of communication in multiprocessor and distributed systems [14]. The *uniform*, *hot-spot* and *local* were called random patterns, because both the source and destination nodes are chosen in a randomly way. In the uniform pattern, all nodes have the same probability of being selected. The hot-spot pattern is similar to uniform. However, for the destination nodes, a particular node has a higher probability of selection. In local pattern, only nodes around the source node can be selected as a destination.

The *complement*, *matrix transpose 1* and *matrix transpose 2* were called deterministic patterns. Although the selection of source nodes is random (following the uniform distribution), the destination nodes are selected according to the position of the source nodes. In the complement pattern, for a source node in the position (x, y), the destination node is in the position $(size - x + 1, size - y + 1)$, where $size$ is the number of nodes in a column or row of the mesh. For patterns matrix transpose 1 and 2, the destination nodes are respectively in the positions $(size - y + 1, size - x + 1)$ and (y, x).

11.7.1.2   Injection Rate

The packet *injection rate* relates the transmission time of flits (from resource to switch) and idle time between the end of the transmission of a packet and the beginning of the transmission of next packet. The injection rate is a fraction of the network channel total bandwidth.

Two different injection rates are shown in Fig. 11.8. It is assumed that each flit is transmitted in one cycle, and that the period between the start of transmission of two consecutive packets is fixed. At the rate of 50%, a 5 flit size packet is injected into the network; only after a idle time of 5 cycles, the next packet starts to be injected. In this situation, only half of the total transmission capacity is used. At the rate of 100%, there is no idle time between packets. The use of injection rates lower than 100% is interesting in situations of network congestion, since the late flits can be sent during idle time between packets.

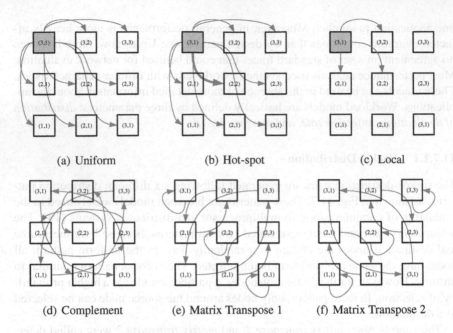

Fig. 11.7 Possible communication pairs in a 3×3 mesh

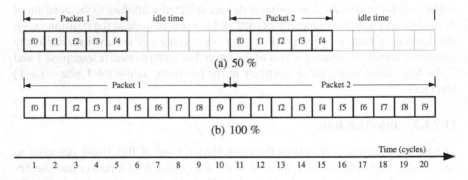

Fig. 11.8 Two different injection rates

11.7.1.3 Packet Size

The packet size can also be shaped in various ways. Two values must be distinguished: the size of a packet and its amount of flits. The size L of a package is the value of its total length in bits. In turn, the amount of flits defined by Equation 11.10 is the largest integer value obtained by dividing the packet size by W, the size of a *phit* (physical unit, width of channel bits).

$$\#flits = \left\lceil \frac{L}{W} \right\rceil \tag{11.10}$$

Generally, packet length is defined as a constant in simulations [14]. Alternatively, the length may be made variable in simulations, when studying the effects of different packet sizes on network. In this situation, the size can be chosen at random according to a specific probability distribution, such as the spatial distribution of packets.

In this work, the amount of flits is associated with the injection rate. The number of cycles from the start of transmission of a packet and the start of the next is defined as the fixed value of 20 cycles. Thus, the amount of flits varies according to desired injection rate as seen in Table 11.2.

Table 11.2 Amount of flits in each injection rate

Injection rate	Flits	Idle cycles
10%	02	18
20%	04	16
30%	06	14
40%	08	12
50%	10	10
60%	12	08
70%	14	06
80%	16	04
90%	18	02
100%	20	00

11.7.1.4 Simulation Results

For each simulation, we obtained the total latency and the average latency per packet. The total latency is the sum of the individual latency of all packets being transmitted on the network. The individual latency is the amount of simulation cycles that have elapsed since the injection of the first flit of a packet until the beginning of injection of the next packet of the same message. The average latency is the latency value obtained divided by the total number of packets.

Results are shown in the Fig. 11.9. The general purpose of these tests is to verify the variation of latency under different injection rates. The curves of *latency/packet × injection rate* are, in fact, a mean of the values obtained for different quantity of packets. Each graph illustrates these curves for the four routing algorithms adopted.

The latency values (obtained in simulations) can be also arranged as function of number of packets. In Fig. 11.10, for each traffic pattern, *latency/packet × number of packets* curves are shown. For these curves, the value *latency/packet* is the average (of obtained values for a same number of packets), for different injection rates.

For all traffic patterns, the latency curve of REAS is located below the curves of the other methods, indicating its ability to search for routes that provide a shorter transmission time. This performance is slightly better than the others at low injection

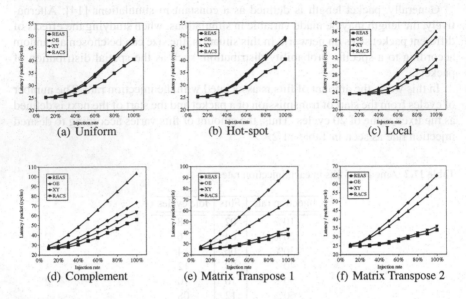

Fig. 11.9 Results for the network under six different traffic patterns

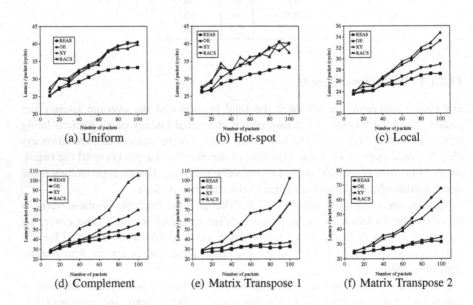

Fig. 11.10 Results for the network under six different traffic patterns

rates, becoming more evident in rates above 50%. The RACS has a latency curve similar to the obtained by XY and OE algorithms for the uniform and hot-spot traffic patterns. For Local and Complement patterns these curves differ, with RACS getting lower latency values compared to the XY and OE. In matrix transpose patterns, the RACS achieved similar results to those obtained by REAS.

11.7.2 Simulation with Synthetic Task Graphs

In these simulations, we used five sets of graphs of synthetic applications. These graphs were randomly generated with the aid of the software *Task Graph For Free* (TGFF) [9]. The TGFF is a general purpose, user controllable pseudo-random graph generator, widely used in embedded real-time systems research. The software works based on a script, where the user defines parameters such as total number of tasks, levels, or tasks per level. A example of graph created by TGFF is shown in Fig. 11.11, with two intermediary levels, each one with three tasks.

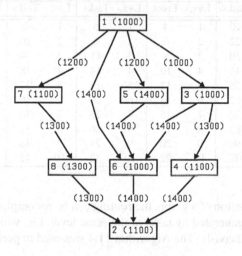

Fig. 11.11 Example of TGFF generated graph

The task graph is composed by a set of nodes (rectangles) and arcs (arrows). The numbers in each node is a task index, where "1" is the start task and "2" is the end task, and the execution time of the task, in cycles. The values in the arcs are the number of bits of the transmitted packet. In orther to explore the behavior of applications with parallel characteristics, tasks were generated following a *fork-join* structure, with the start task sending packets to several destinations, and the end task receiving packets from several origins. Between start and end tasks exist intermediate tasks, arranged in various levels of parallelism. Tasks at the same level can run concurrently and independently.

Therefore, 50 graphs are generated, being arranged in 5 sets of 10 graphs. Each set, called $Ex1$, $Ex2$, $Ex3$, $Ex4$ and $Ex5$, has a different characteristic on the maximum number of tasks. Within a set, each of the ten tasks are differentiated by the number of intermediate levels.

In the $Ex1$ set, graphs are generated based only in the total number of tasks. The structure of nodes and arcs are build in a randomly way, with only restriction the format of fork-join graph. Thus, the graph $Ex1.1$ (the first of set $Ex1$) is composed of ten tasks, while $Ex1.10$ (the last) is composed of a total of one hundred tasks. In the other four sets, the structure of each application graph is based mainly in the number of tasks per level and the number of levels. Applications in $Ex2$ have two tasks per level - the set $Ex3$ have three tasks, and so on. In each graph from $Ex2.1$ to $Ex2.10$, the number of levels vary from one to ten intermediary levels. Table 11.3 shows the number of levels and the total number of tasks in all five sets of graphs.

Table 11.3 Number of intermediary levels and tasks in all application graphs

	$Ex1$		$Ex2$		$Ex3$		$Ex4$		$Ex5$	
	Lvs.	Tasks	Lvs.	Tasks	Lvs.	Tasks	Lvs.	Tasks	Lvs.	Tasks
1	6	10	1	4	1	5	1	6	1	7
2	8	22	2	6	2	8	2	10	2	12
3	10	30	3	8	3	11	3	14	3	17
4	12	42	4	10	4	14	4	18	4	22
5	14	50	5	12	5	17	5	22	5	27
6	16	62	6	14	6	20	6	26	6	32
7	16	70	7	16	7	23	7	30	7	37
8	16	82	8	18	8	26	8	34	8	42
9	18	90	9	20	9	29	9	38	9	47
10	22	102	10	22	10	32	10	42	10	52

From the information of a graph, the routing can be accomplished by identifying which packets are generated by tasks at the same level, i.e., which packets may be transmitted simultaneously. The Algorithm 11.4 was used to perform this process.

11.7.2.1 Simulation Results

The simulations were performed by submitting applications to four different routing algorithms and measuring its total execution time. This consists of execution of all individual tasks on a critical path plus the communication time of these tasks. The so-called *packet delay* is the difference between the value obtained using a specific routing algorithm and the optimal value of the network without congestion. To calculate this ideal value, we used a modified XY algorithm, called *dummy XY*. In this routing, the XY algorithm is used to define the communication time using shortest paths. But unlike the real XY (and any other routing algorithm), the potential congestion delays are not counted.

Algorithm 11.4. Mapping and routing of application

Require: Task Graph;
 1: **define** size of NoC;
 2: perform the mapping;
 3: **for** $l = 1 \rightarrow \#levels$ **do**
 4:    get all arcs in level l;
 5:    read t_{start} of source tasks;
 6:    perform the routing;
 7:    write t_{start} of destination tasks;
 8: **end for**
 9: $t_{execution} \leftarrow t_{start}(last\ task) + t_{comp}(last\ task)$
10: **return** routing paths, $t_{execution}$;

The Fig. 11.12 shown results of the performed simulations. The values of packet delay is presented for the four routing algorithms in each of 50 applications. These values are a mean of packet delay in 10 different mappings.

The results show the REAS getting the best results in the simulations when compared with other routing algorithms. The REAS is exceeded only in 12 of 50 tests. These low delay values show that the REAS is able to find good solutions to routing problem, independent of mapping or complexity of graph. For routing based on XY algorithm and Odd-Even turn model, there is a wide variation in average delays obtained for a given set of graphs. This large deviation in the delay values may suggest that XY and OE are very sensitive to mapping adopted, even more than the complexity of the application. The worst results were obtained with the second proposed routing. The values found by RACS are increasing due to the complexity of the used graphs.

11.7.2.2 Statistical Analysis

A given set of statistical results have significance if it is unlikely that these have occurred by chance. The presented results show the REAS being able to get better results than other routing algorithms used for comparison. In order to determine whether the results obtained by REAS are significantly better than those obtained by RACS, XY and OE, we performed a statistical test of significance. The most commonly used method of comparing proportions uses the χ^2-test [7]. This test makes it possible to determine whether the difference existing between two groups of data is significant or just a chance occurrence.

$$E_{at} = \frac{\sum\limits_{x \in A} O_{xt} \cdot \sum\limits_{y \in T} O_{ay}}{\sum\limits_{(x,y) \in A \times T} O_{x,y}} \tag{11.11}$$

For the sake of completeness, we explain briefly how the test works. χ^2-test determines the differences between the observed and expected measures. The observed

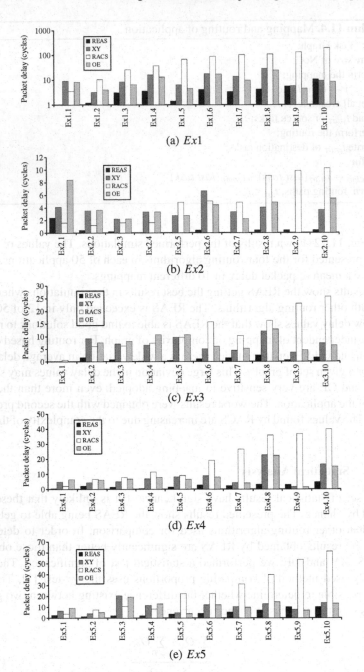

Fig. 11.12 Packet delay in 5 sets of applications

values are the actual experimental results, whereas the expected ones refer to the hypothetical distribution based on the overall proportions between the two compared algorithms if these are alike. Tests were conducted to $REAS \times XY$, $REAS \times OE$ and $REAS \times RACS$ separately for each set of simulations ($Ex1$ to $Ex5$) using the obtained values of latency. The calculation of the expected value is performed with (11.11). The value of χ^2 is calculated with (11.12), where A and T are the sets of simulations and algorithms used in each test. The value O_{at} is the observed latency in the simulation t by the routing algorithm a.

$$\chi^2 = \sum_{(a,t) \in A \times T} \frac{(O_{at} - E_{at})^2}{E_{at}} \tag{11.12}$$

The computed values for χ^2 for each of the comparisons are given in Table 11.4. The use of the χ^2-test is recommended when the proportions are small. Therefore, the time quantities were converted to 0.1 *cycle* instead of 1 *cycle*, thus avoiding the limitation imposed for the usage of the test.

Table 11.4 Significance levels obtained with χ^2 test

		Degrees of freedom	χ^2	critical χ^2	Significance
	$REAS \times XY$	8	155,5	26,12	99,90%
$Ex1$	$REAS \times OE$	8	124,48	26,12	99,90%
	$REAS \times RACS$	8	68,38	26,12	99,90%
	$REAS \times XY$	5	32,94	20,52	99,90%
$Ex2$	$REAS \times OE$	5	13,25	12,83	99,90%
	$REAS \times RACS$	6	120,12	22,46	97,50%
	$REAS \times XY$	8	24,43	24,35	99,80%
$Ex3$	$REAS \times OE$	8	33,47	26,12	99,90%
	$REAS \times RACS$	8	105,25	26,12	99,90%
	$REAS \times XY$	9	90,93	27,88	99,90%
$Ex4$	$REAS \times OE$	9	99,36	27,88	99,90%
	$REAS \times RACS$	9	37,57	27,88	99,90%
	$REAS \times XY$	9	144,53	27,88	99,90%
$Ex5$	$REAS \times OE$	9	166,6	27,88	99,90%
	$REAS \times RACS$	9	31,21	27,88	99,90%

The critical value of χ^2 is 0.05 (i.e. 95% of confidence) and considered the limit to assume the tested hypothesis. The degree of freedom depends on the amount of results used to compute χ^2. Assuming that the results are organized in a two-dimensional array of r rows and c columns, the degree of freedom is defined by $(r-1) \times (c-1)$.

As can be seen in the Table 11.4, null hypothesis may be discarded in all cases. Based on this statistical analysis, the REAS can be considered significantly better than the algorithms XY, OE and RACS.

11.7.3 Simulation with Real World Applications

The simulations were performed by submitting applications to four different routing algorithms and measuring its total execution time. This consists of execution of all individual tasks on a critical path plus the communication time of these tasks. The so-called *packet delay* is the difference between the value obtained using a specific routing algorithm and the optimal value of the network without congestion. To calculate this ideal value, we used a modified XY algorithm, called *dummy XY*. In this routing, the XY algorithm is used to define the communication time using shortest paths. But unlike the real XY (and any other routing algorithm), the potential congestion delays are not counted.

11.7.3.1 Applications from E3S

The *Embedded Systems Synthesis benchmarks Suite* (E3S) [8] is a collection of task graphs , representing real applications based on embedded processors from Embedded Microprocessor Consortium (EEMBC). It was developed to be used in system-level allocation, assignment and scheduling research. The E3S contains the characteristics of 17 embedded processors. These processors are characterized by the measured execution times of 47 different type of tasks, power consumption derived from processor data sheets, and additional information, such as die size, price, clock frequency and power consumption during idle state. In addition, E3S contains task graphs of common tasks in auto-industry, networking, telecommunication and office automation. Each one of the nodes of these task graphs is associated with a task type. A task type is a processor instruction or a set of instructions, e.g., FFT, inverse FFT, floating point operation, etc.

In this study, we used 16 graphs found in E3S, which represent serial and parallel applications. Information about the number of levels and tasks for each application is shown in Table 11.5. The AMD-ElanSC520 was selected, which is able to perform all 47 tasks.

11.7.3.2 SegImag Application

Another application used in this work is the segmentation of images for object recognition, *SegImag* [21]. This application aims to accelerate the process of identifying the number of objects in an image. For this purpose, the original image must be split in parts, where each segment is handled by an auxiliary processor. In addition, the SegImag contains two other processing elements: a central processor, which receives the results of each auxiliary processor, and an external memory which stores the image to be segmented.

In the original implementation of SegImag, the amount of auxiliary processors is parameterized. The segmentation is directly related to the amount of processing elements, since each segment must be processed by a auxiliary processor. In the present study, we used the implementation shown in SegImag [4], which

Table 11.5 List of applications in E3S

label	Application Name	# Tasks	# Levels
TG1	auto-indust-tg0	6	6
TG2	auto-indust-tg1	4	4
TG3	auto-indust-tg2	9	8
TG4	auto-indust-tg3	5	5
TG5	consumer-tg0	7	5
TG6	consumer-tg1	5	4
TG7	networking-tg1	4	4
TG8	networking-tg2	4	4
TG9	networking-tg3	4	4
TG10	office-tg0	5	4
TG11	telecom-tg0	4	4
TG12	telecom-tg1	6	5
TG13	telecom-tg2	6	5
TG14	telecom-tg3	3	3
TG15	telecom-tg4	3	3
TG16	telecom-tg5	2	2

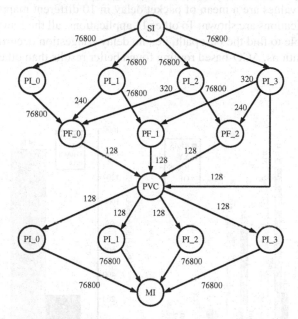

Fig. 11.13 Task graph of SegImage application

represents a specific application to the image segmented into four parts. The task graph of this application is shown in Fig. 11.13. The values in each arc represent the amount of bits transmitted between tasks, considering an image with 640×480 pixels. The run-time values were based on allocation results presented in [4], where each task in SegImag task graph was associated with a task performed by a processor repository E3S. This information concerning the task data is organized in Table 11.6. As the time unit of tasks is in *ns*, in this test we considered that the *simulator cycle* is equal to 1 *ns*.

Table 11.6 Tasks of E3S used in SegImag application

label	Task Name	id	Task Time (ns)	Proc. Name
SI	Decompress JPEG	455	7×10^7	ST20C2
PI	Fixed Point Complex FFT - Data3 (sine)	449	$1,2 \times 10^5$	ST20C2
PF	Basic floating point	371	$8,9 \times 10^2$	MPC555
PCV	Autocorrelation - Data2 (sine)	439	$6,9 \times 10^4$	ST20C2
MI	Compress JPEG	454	$8,7 \times 10^7$	ST20C2

11.7.3.3 Simulation Results

The Fig. 11.14 shown results of simulations with real world applications. The values of packet delay is presented for the four routing algorithms - XY, OE, REAS and RACS. These values are a mean of packet delay in 10 different mappings. Results of only 5 applications are shown. To other 11 applications, all the four routing algorithms were able to find the best path, i.e., no delay congestion occurred. In four of the five applications, ACO-based routing found better results than other algorithms.

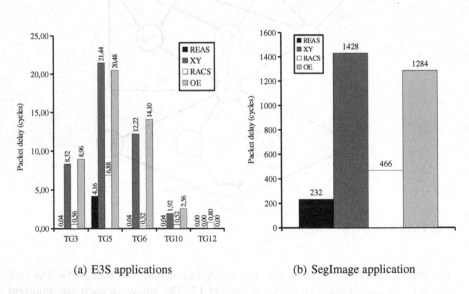

(a) E3S applications　　　　　　　　(b) SegImage application

Fig. 11.14 Results for simulation with applications

11.8 Summary

Static routing is an efficient solution in NoCs designed to run always the same set of applications, since communication paths need be defined only one time. In this chapter we propose the use of ACO-based algorithms in the optimization of communication paths in the static routing step in NoC design. The performance of these algorithms was evaluated in three different approaches: analyzing the behavior of network under different random traffic patterns; using several synthetic application task graphs; and using task graphs of real world applications. Best results were obtained with REAS algorithm. Future work may be conducted in two directions. First, it is of interest to determine how to enhance the performance of the proposed methods, and the study of the use of other ant algorithms. Second, it is still necessary to perform an analysis covering both routing, mapping and allocation, so that these tasks can be used in future in an EDA tool, thus assisting in the design of systems based on the NoC platform.

References

1. Benini, L., De Micheli, G.: Networks on chips: A new soc paradigm. Computer 35(1), 70–78 (2002)
2. Bonabeau, E., Dorigo, M., Theraulaz, G.: Swarm intelligence: from natural to artificial systems. Oxford University Press, USA (1999)
3. Chiu, G.M.: The odd-even turn model for adaptive routing. IEEE Transactions on Parallel and Distributed Systems 11(7), 729–738 (2000)
4. da Silva, M.V.C., Nedjah, N., Mourelle, L.M.: Efficient mapping of an image processing application for a network-on-chip based implementation. International Journal of High Performance Systems Architecture 2(1), 46–57 (2009)
5. da Silva, M.V.C., Nedjah, N., Mourelle, L.M.: Optimal ip assignment for efficient noc-based system implementation using nsga-ii and microga. IJCIS 2(2), 115–123 (2009)
6. da Silva, M.V.C., Nedjah, N., Mourelle, L.M.: Power-aware multi-objective evolutionary optimisation for application mapping on network-on-chip platforms. International Journal of Electronics 97(10), 1163–1179 (2010)
7. Diaconis, P., Efron, B.: Testing for independence in a two-way table: new interpretations of the chi-square statistic. The Annals of Statistics 13(3), 845–874 (1985)
8. Dick, R.: Embedded system synthesis benchmarks suites (E3S),
   http://ziyang.eecs.umich.edu/~dickrp/e3s/ (accessed May 2, 2012)
9. Dick, R.P., Rhodes, D.L., Wolf, W.: Tgff: task graphs for free. In: Proceedings of the 6th International Workshop on Hardware/Software Codesign, pp. 97–101. IEEE Computer Society (1998)
10. Dorigo, M., Birattari, M., Stutzle, T.: Ant colony optimization. IEEE Computational Intelligence Magazine 1(4), 28–39 (2006)
11. Dorigo, M., Gambardella, L.M.: Ant colony system: A cooperative learning approach to the traveling salesman problem. IEEE Transactions on Evolutionary Computation 1(1), 53–66 (1997)
12. Dorigo, M., Maniezzo, V., Colorni, A.: Ant system: optimization by a colony of cooperating agents. IEEE Transactions on Systems, Man, and Cybernetics, Part B: Cybernetics 26(1), 29–41 (1996)

13. Duato, J.: A new theory of deadlock-free adaptive routing in wormhole networks. IEEE Transactions on Parallel and Distributed Systems 4(12), 1320–1331 (1993)
14. Duato, J., Yalamanchili, S., Ni, L.M.: Interconnection networks: An engineering approach. Morgan Kaufmann (2003)
15. Edwards, S., Lavagno, L., Lee, E.A., Sangiovanni-Vincentelli, A.: Design of embedded systems: Formal models, validation, and synthesis. Proceedings of the IEEE 85(3), 366–390 (1997)
16. Esser, R., Knecht, R.: Intel paragon xp/s-architecture and software enviroment. In: Anwendungen, Architekturen, Trends, Seminar, pp. 121–141. Springer (1993)
17. Glass, C.J., Ni, L.M.: The turn model for adaptive routing. In: SIGARCH Computer Architecture News, vol. 20, pp. 278–287. ACM (1992)
18. Goss, S., Aron, S., Deneubourg, J., Pasteels, J.: Self-organized shortcuts in the argentine ant. Naturwissenschaften 76, 579–581 (1989), doi:10.1007/BF00462870
19. A. Intel. Touchstone delta system description. Supercomputer Systems Division, Intel Corporation, Beaverton, OR, 97006 (1991)
20. Jóźwiak, L., Nedjah, N., Figueroa, M.: Modern development methods and tools for embedded reconfigurable systems: A survey. Integration, The VLSI Journal 43(1), 1–33 (2010)
21. Marcon, C.A.M.: Modelos para o Mapeamento de Aplicações em Infra-estruturas de Comunicação Intrachip. PhD thesis, Universidade Federal do Rio Grande do Sul (2005)
22. Moraes, F., Calazans, N., Mello, A., Moller, L., Ost, L.: Hermes: an infrastructure for low area overhead packet-switching networks on chip. Integration, The VLSI Journal 38(1), 69–93 (2004)
23. Mourelle, L.M., Ferreira, R.E., Nedjah, N.: Migration selection of strategies for parallel genetic algorithms: implementation on networks on chips. International Journal of Electronics 97(10), 1227–1240 (2010)
24. Nedjah, N., Da Silva, M.V.C., Mourelle, L.M.: Customized computer-aided application mapping on noc infrastructure using multi-objective optimization. Journal of Systems Architecture: The EUROMICRO Journal 57(1), 79–94 (2011)
25. Nedjah, N., da Silva, M.V.C., Mourelle, L.M.: Preference-based multi-objective evolutionary algorithms for power-aware application mapping on noc platforms. Expert Systems with Applications: An International Journal 39(3), 2771–2782 (2012)
26. Ni, L.M., McKinley, P.K.: A survey of wormhole routing techniques in direct networks. Computer 26(2), 62–76 (1993)
27. Seitz, C.L., Athas, W.C., Flaig, C.M., Martin, A.J., Seizovic, J., Steele, C.S., Su, W.K.: The architecture and programming of the ametek series 2010 multicomputer. In: Proceedings of the Third Conference on Hypercube Concurrent Computers and Applications: Architecture, Software, Computer Systems, and General Issues, vol. 1, pp. 33–37. ACM (1988)
28. Seitz, C.L., Boden, N.J., Seizovic, J., Su, W.K.: The design of the caltech mosaic c multicomputer. Computer 256, 80 (1993)
29. Zeferino, C.A., Susin, A.A.: Socin: a parametric and scalable network-on-chip. In: Proceedings of the 16th Symposium on Integrated Circuits and Systems Design, SBCCI 2003, pp. 169–174. IEEE (2003)

Printed in the United States
By Bookmasters